Lecture Notes
in Control and Information Sciences 276

Editors: M. Thoma · M. Morari

Springer

Berlin
Heidelberg
New York
Barcelona
Hong Kong
London
Milan
Paris
Tokyo

Engineering ONLINE LIBRARY

http://www.springer.de/engine/

Z. Bubnicki

Uncertain Logics, Variables and Systems

With 36 Figures

 Springer

Author

Prof. Zdzislaw Bubnicki
Wroclaw University of Technology
Institute of Control and Systems Engineering
Wyb. Wyspianskiego 27
50-370 Wroclaw, Poland

Cataloging-in-Publication Data applied for
Die Deutsche Bibliothek – CIP-Einheitsaufnahme
Bubnicki, Zdzislaw:
Uncertain logics, variables and systems / Z. Bubnicki. - Berlin ; Heidelberg ;
New York ; Barcelona ; Hong Kong ; London ; Milan ; Paris ; Tokyo :
Springer, 2002
　(Lecture notes in control and information sciences ; 276)
　(Engineering online library)
　ISBN 3-540-43235-3

ISBN 3-540-43235-3　Springer-Verlag Berlin Heidelberg New York

Springer-Verlag Berlin Heidelberg New York
a member of BertelsmannSpringer Science + Business Media GmbH

http://www.springer.de

© Springer-Verlag Berlin Heidelberg 2002
Printed in Germany

Typesetting: Digital data supplied by author. Data-conversion by PTP-Berlin, Stefan Sossna e.K.
Cover-Design: design & production GmbH, Heidelberg
Printed on acid-free paper　　SPIN 10867682　　62/3020Rw - 5 4 3 2 1 0

Foreword

The ideas of uncertain variables based on uncertain logics have been introduced and developed for a wide class of uncertain systems. The purpose of this monograph is to present basic concepts, definitions and results concerning the uncertain variables and their applications to analysis and decision problems in uncertain systems described by traditional mathematical models and by knowledge representations.

I hope that the book can be useful for graduate students, researchers and all readers working in the field of control and information science. Especially for those interested in the problems of uncertain decision support systems and uncertain control systems.

I wish to express my gratitude to my co-workers from the Institute of Control and Systems Engineering of Wroclaw University of Technology, who assisted in the preparation of the manuscript. My special thanks go to Dr L. Siwek for the valuable remarks and for his work concerning the formatting of the text.

This work was supported in part by the Polish Committee for Scientific Research under the grants no. 8 T11A 022 14 and 8 T11C 012 16.

Preface

Uncertainty is one of the main features of complex and intelligent decision making systems. There exists a great variety of definitions and descriptions of uncertainties and uncertain systems. The most popular non-probabilistic approaches are based on fuzzy sets theory and related formalisms such as evidence and possibility theory (e.g. [1, 2, 37-42, 51-55]). The different formulations of decision making problems and various proposals for reasoning under uncertainty are adequate to the different formal models of uncertainty. Special approaches have been presented for uncertainty in expert systems [47] and for uncertain control systems (e.g. [21, 44, 50]). This work concerns a class of uncertain systems containing unknown parameters in their mathematical descriptions: in traditional mathematical models or in knowledge representations in the knowledge-based systems. For such systems a concept of so called uncertain variables and its application to the analysis and decision making problems have been developed [15, 16, 17, 20, 26, 28-36]. The purpose of this work is to present a basic theory of the uncertain variables and a unified description of their applications in the different cases of the uncertain systems.

In the traditional case, for a static system described by a function $y = \Phi(u, x)$ where u, y, x are input, output and parameter vectors, respectively, the decision problem may be formulated as follows: to find the decision u^* such that $y = y^*$ (the desirable output value). The decision u^* may be obtained for the known Φ and x. Let us now assume that x is unknown. In the probabilistic approach x is assumed to be a value of a random variable $\tilde{x}$ described by the probability distribution. In our approach the unknown parameter x is a value of an uncertain variable $\bar{x}$ for which an expert gives the certainty distribution $h(x) = v(\bar{x} \cong x)$ where v denotes a certainty index of the soft property: "$\bar{x}$ is approximately equal to x" or "x is the approximate value of $\bar{x}$". The uncertain variables, related to random variables and fuzzy numbers, are described by the set of values X and their certainty distributions which correspond to probability distributions for the random variables and to membership functions for the fuzzy numbers. To define the uncertain variable, it is necessary to give $h(x)$ and to determine the certainty indexes of the following soft properties: "$\bar{x} \,\tilde{\in}\, D_x$" for $D_x \subset X$ which means: "the approximate value of $\bar{x}$ belongs to D_x" or "$\bar{x}$ belongs approximately to D_x", and "$\bar{x} \,\tilde{\notin}\, D_x$" = "$\neg(\bar{x} \,\tilde{\in}\, D_x)$" which means "$\bar{x}$ does not belong approximately to D_x". To determine the certainty indexes for the properties:

$\neg(\bar{x} \tilde{\in} D_x)$, $(\bar{x} \tilde{\in} D_1) \vee (\bar{x} \tilde{\in} D_2)$ and $(\bar{x} \tilde{\in} D_1) \wedge (\bar{x} \tilde{\in} D_2)$ where $D_1, D_2 \subseteq X$, it is necessary to introduce an *uncertain logic* which deals with the soft predicates of the type "$\bar{x} \tilde{\in} D_x$". Four versions of the uncertain logic have been introduced (Sects. 2.1 and 2.2) and then two of them have been used for the formulation of two versions of the uncertain variable (Sects. 2.3 and 2.4).

For the proper interpretation (semantics) of these formalisms it is convenient to consider $\bar{x} = g(\omega)$ as a value assigned to an element $\omega \in \Omega$ (a universal set). For fixed ω its value $\bar{x}$ is determined and $\bar{x} \in D_x$ is a crisp property. The property $\bar{x} \tilde{\in} D_x = x \in D_x = $ "the approximate value of $\bar{x}$ belongs to D_x" is a soft property because $\bar{x}$ is unknown and the evaluation of "$\bar{x} \tilde{\in} D_x$" is based on the evaluation of $\bar{x} \cong x$ for the different $x \in X$ given by an expert. In the first version of the uncertain variable $v(\bar{x} \tilde{\in} D_x) \neq v(\bar{x} \tilde{\notin} \overline{D}_x)$ where $\overline{D}_x = X - D_x$ is the complement of D_x. In the second version called C-uncertain variable $v_c(\bar{x} \tilde{\notin} D_x) = v_c(\bar{x} \tilde{\in} \overline{D}_x)$ where v_c is the certainty index in this version: $v_c(\bar{x} \tilde{\in} D_x) = \frac{1}{2}[v(\bar{x} \tilde{\in} D_x) + v(\bar{x} \tilde{\notin} \overline{D}_x)]$. The uncertain variable in the first version may be considered as a special case of the possibilistic number with a specific interpretation of $h(x)$ described above. In our approach we use soft properties of the type "P is approximately satisfied" where P is a crisp property, in particular $P = "\bar{x} \in D_x"$. It allows us to accept the difference between $\bar{x} \tilde{\in} D_x$ and $\bar{x} \tilde{\notin} \overline{D}_x$ in the first version. More details concerning the relations to random variables and fuzzy numbers are given in Chap. 6. Now let us pay attention to the following aspects which will be more clear after the presentation of the formalisms and semantics in Chap. 1:

1. To compare the meanings and practical utilities of different formalisms, it is necessary to take into account their semantics. It is specially important in our approach. The definitions of the uncertain logics and consequently the uncertain variables contain not only the formal description but also their interpretation. In particular, the uncertain logic may be considered as special cases of multi-valued predicate logic with a specific semantics of the predicates. It is worth noting that from the formal point of view the probabilistic measure is a special case of the fuzzy measure and the probability distribution is a special case of the membership function in the formal definition of the fuzzy number when the meaning of the membership function is not described.

2. Even if the uncertain variable in the first version may be formally considered as a very special case of the fuzzy number, for the simplicity and the unification it is better to introduce it independently (as has been done in the work) and not as a special case of the much more complicated formalism with different semantics and applications.

3. The *uncertainty* is understood here in a narrow sense of the word and concerns an incomplete or imperfect knowledge of something which is necessary to

solve the problem. In our considerations it is the knowledge on the parameters in the mathematical model of the decision making problem, and is related to a fixed expert who gives the description of the uncertainty.

4. In the majority of interpretations the value of the membership function means a *degree of truth* of a soft property determining the fuzzy set. In our approach, " $\bar{x} \in D_x$ " and " $x \in D_x$ " are crisp properties, the soft property " $\bar{x} \tilde{\in} D_x$ " is introduced because the value of $\bar{x}$ is unknown and $h(x)$ is a *degree of certainty* (or $1 - h(x)$ is a degree of uncertainty).

In Chaps. 2–5 the application of the uncertain variables to basic analysis and decision making problems is presented for the systems with the different forms of the mathematical descriptions. Additional considerations concerning special and related problems are presented in Chap. 7.

Contents

1 Uncertain Logics and Variables

1.1 Uncertain Logic

Our considerations are based on multi-valued logic. To introduce terminology and notation employed in our presentation of uncertain logic and uncertain variables, let us remind that multi-valued (exactly speaking – infinite-valued) propositional logic deals with propositions $(\alpha_1, \alpha_2, ...)$ whose logic values $w(\alpha) \in [0, 1]$ and

$$w(\neg\alpha) = 1 - w(\alpha),$$

$$w(\alpha_1 \vee \alpha_2) = \max\{w(\alpha_1), w(\alpha_2)\}, \qquad (1.1)$$

$$w(\alpha_1 \wedge \alpha_2) = \min\{w(\alpha_1), w(\alpha_2)\}.$$

Multi-valued predicate logic deals with predicates $P(x)$ defined on a set X, i.e. properties concerning x, which for the fixed value of x form propositions in multi-valued propositional logic, i.e.

$$w[P(x)] \overset{\Delta}{=} \mu_p(x) \in [0, 1] \quad \text{for each } x \in X. \qquad (1.2)$$

For the fixed x, $\mu_p(x)$ denotes *degree of truth*, i.e. the value $\mu_p(x)$ shows to what degree P is satisfied. If for each $x \in X$ $\mu_p(x) \in \{0, 1\}$ then $P(x)$ will be called here a *crisp* or a *well-defined property*, and $P(x)$ which is not well-defined will be called a *soft property*. The crisp property defines a set

$$D_x = \{x \in X: w[P(x)] = 1\} \overset{\Delta}{=} \{x \in X: P(x)\}. \qquad (1.3)$$

Consider now a universal set Ω, $\omega \in \Omega$, a set X which is assumed to be a metric space, a function $g: \Omega \rightarrow X$, and a crisp property $P(x)$ in the set X. The property P and the function g generate the crisp property $\Psi(\omega, P)$ in Ω: "For the value $\bar{x} = g(\omega) \overset{\Delta}{=} \bar{x}(\omega)$ assigned to ω the property P is satisfied", i.e.

$$\Psi(\omega, P) = P[\bar{x}(\omega)].$$

Let us introduce now the property $G(\bar{x}, x) = "\bar{x} \cong x"$ for $\bar{x}, x \in X$, which means: "$\bar{x}$ is approximately equal to x". The equivalent formulations are: "x is the approximate value of $\bar{x}$" or "x belongs to a small neighbourhood of $\bar{x}$" or "the value of the metric $d(x, \bar{x})$ is small". Note that $G(\bar{x}, x)$ is a reflexive, symmetric and transitive relation in $X \times X$. For the fixed ω, $G[\bar{x}(\omega), x] \triangleq G_\omega(x)$ is a soft property in X. The properties $P(x)$ and $G_\omega(x)$ generate the soft property $\overline{\Psi}(\omega, P)$ in Ω: "the approximate value of $\bar{x}(\omega)$ satisfies P" or "$\bar{x}(\omega)$ approximately satisfies P", i.e.

$$\overline{\Psi}(\omega, P) = G_\omega(x) \wedge P(x) = [\bar{x}(\omega) \cong x] \wedge P(x) \tag{1.4}$$

where x is a free variable. The property $\overline{\Psi}$ may be denoted by

$$\overline{\Psi}(\omega, P) = "\bar{x}(\omega) \,\tilde{\in}\, D_x" \tag{1.5}$$

where D_x is defined by (1.3) and "$\bar{x} \,\tilde{\in}\, D_x$" means: "the approximate value of $\bar{x}$ belongs to D_x" or "$\bar{x}$ approximately belongs to D_x". Denote by $h_\omega(x)$ the logic value of $G_\omega(x)$:

$$w[G_\omega(x)] \triangleq h_\omega(x), \qquad \bigwedge_{x \in X} (h_\omega(x) \geq 0), \tag{1.6}$$

$$\max_{x \in X} h_\omega(x) = 1. \tag{1.7}$$

Definition 1.1 (uncertain logic): The uncertain logic is defined by a universal set Ω, a metric space X, crisp properties (predicates) $P(x)$, the properties $G_\omega(x)$ and the corresponding functions (1.6) for $\omega \in \Omega$. In this logic we consider soft properties (1.4) generated by P and G_ω. The logic value of $\overline{\Psi}$ is defined in the following way

$$w[\overline{\Psi}(\omega, P)] \triangleq v[\overline{\Psi}(\omega, P)] = \begin{cases} \max\limits_{x \in D_x} h_\omega(x) & \text{for } D_x \neq \varnothing \\ 0 & \text{for } D_x = \varnothing \end{cases} \tag{1.8}$$

and is called a degree of certainty or *certainty index*. The operations for the certainty indexes are defined as follows:

$$v[\neg \overline{\Psi}(\omega, P)] = 1 - v[\overline{\Psi}(\omega, P)], \tag{1.9}$$

$$v[\Psi_1(\omega, P_1) \vee \Psi_2(\omega, P_2)] = \max\{v[\Psi_1(\omega, P_1)], v[\Psi_2(\omega, P_2)]\}, \tag{1.10}$$

$$v\,[\Psi_1(\omega, P_1) \wedge \Psi_2(\omega, P_2)] = \begin{cases} 0 & \text{if for each } x \;\; w\,(P_1 \wedge P_2) = 0 \\ \min \{v\,[\Psi_1(\omega, P_1)], v\,[\Psi_2(\omega, P_2)]\} & \text{otherwise} \end{cases}$$

(1.11)

where Ψ_1 is $\overline{\Psi}$ or $\neg\overline{\Psi}$, and Ψ_2 is $\overline{\Psi}$ or $\neg\overline{\Psi}$.   □

Using the notation (1.5) we have

$$v\,[\overline{x}(\omega) \,\tilde{\notin}\, D_x] = 1 - v\,[\overline{x}(\omega) \,\tilde{\in}\, D_x],$$

(1.12)

$$v\,[\overline{x}(\omega) \,\tilde{\in}\, D_1 \vee \overline{x}(\omega) \,\tilde{\in}\, D_2] = \max \{v\,[\overline{x}(\omega) \,\tilde{\in}\, D_1], v\,[\overline{x}(\omega) \,\tilde{\in}\, D_2]\},$$ (1.13)

$$v\,[\overline{x}(\omega) \,\tilde{\in}\, D_1 \wedge \overline{x}(\omega) \,\tilde{\in}\, D_2] = \min \{v\,[\overline{x}(\omega) \,\tilde{\in}\, D_1], v\,[\overline{x}(\omega) \,\tilde{\in}\, D_2]\}$$ (1.14)

for $D_1 \cap D_2 \neq \varnothing$ and 0 for $D_1 \cap D_2 = \varnothing$ – where $\tilde{\in}\, D_1$ and $\tilde{\in}\, D_2$ may be replaced by $\tilde{\notin}\, D_1$ and $\tilde{\notin}\, D_2$, respectively.

From (1.7) and (1.8) $v\,[\overline{x} \,\tilde{\in}\, X] = 1$. Let $D_{x,t}$ for $t \in T$ be a family of sets D_x. Then, according to (1.13) and (1.14)

$$v\,[\bigvee_{t \in T} \overline{x}(\omega) \,\tilde{\in}\, D_{x,t}] = \max_{t \in T} v\,[\overline{x}(\omega) \,\tilde{\in}\, D_{x,t}],$$

(1.15)

$$v\,[\bigwedge_{t \in T} \overline{x}(\omega) \,\tilde{\in}\, D_{x,t}] = \min_{t \in T} v\,[\overline{x}(\omega) \,\tilde{\in}\, D_{x,t}].$$

(1.16)

One can note that $G_\omega(x) = "\overline{x}(\omega) \cong x"$ is a special case of $\overline{\Psi}$ for $D_x = \{x\}$ (a singleton) and

$$v\,[\overline{x}(\omega) \cong x] = h_\omega(x), \qquad v\,[\overline{x}(\omega) \ncong x] = 1 - h_\omega(x).$$

(1.17)

According to (1.4), (1.5), (1.17), (1.15), (1.16)

$$v\,[\overline{x}(\omega) \,\tilde{\in}\, D_x] = v\,[\bigvee_{x \in D_x} \overline{x}(\omega) \cong x] = \max_{x \in D_x} h_\omega(x)$$

which coincides with (1.8), and

$$v\,[\overline{x}(\omega) \,\tilde{\notin}\, D_x] = v\,[\bigwedge_{x \in D_x} \overline{x}(\omega) \ncong x] = \min_{x \in D_x} [1 - h_\omega(x)] = 1 - \max_{x \in D_x} h_\omega(x)$$

which coincides with (1.8) and (1.12). From (1.8) one can immediately deliver the following property: if $P_1 \rightarrow P_2$ for each x (i.e. $D_1 \subseteq D_2$) then

$$v\,[\overline{\Psi}(\omega, P_1)] \leq v\,[\overline{\Psi}(\omega, P_2)] \quad \text{or} \quad v\,[\overline{x}(\omega) \,\tilde{\in}\, D_1] \leq v\,[\overline{x}(\omega) \,\tilde{\in}\, D_2].$$ (1.18)

Theorem 1.1:

$$v [\overline{\varPsi} (\omega, P_1 \vee P_2)] = v [\overline{\varPsi} (\omega, P_1) \vee \overline{\varPsi} (\omega, P_2)], \tag{1.19}$$

$$v [\overline{\varPsi} (\omega, P_1 \wedge P_2)] \leq \min\{v [\overline{\varPsi} (\omega, P_1)], v [\overline{\varPsi} (\omega, P_2)]\}. \tag{1.20}$$

Proof: From (1.8) and (1.10)

$$v [\overline{\varPsi} (\omega, P_1) \vee \overline{\varPsi} (\omega, P_2)] = \max \{\max_{x \in D_1} h_\omega(x), \; \max_{x \in D_2} h_\omega(x)\}$$

$$= \max_{x \in D_1 \cup D_2} h_\omega(x) = v [\overline{\varPsi} (\omega, P_1 \vee P_2)].$$

Inequality (1.20) follows immediately from $D_1 \cap D_2 \subseteq D_1$, $D_1 \cap D_2 \subseteq D_2$ and (1.18). $\qquad \qquad \square$

Theorem 1.2:

$$v [\overline{\varPsi} (\omega, \neg P)] \geq v [\neg \overline{\varPsi} (\omega, P)]. \tag{1.21}$$

Proof: Let $P_1 = P$ and $P_2 = \neg P$ in (1.19). Since $w (P \vee \neg P) = 1$ for each x ($D_x = X$ in this case),

$$1 = v [\overline{\varPsi} (\omega, P) \vee \overline{\varPsi} (\omega, \neg P)] = \max\{v [\overline{\varPsi} (\omega, P)], \; v [\overline{\varPsi} (\omega, \neg P)]\}$$

and

$$v [\overline{\varPsi} (\omega, \neg P)] \geq 1 - v [\overline{\varPsi} (\omega, P)] = v [\neg \overline{\varPsi} (\omega, P)]. \qquad \square$$

Inequality (1.21) may be written in the form

$$v [\overline{x}(\omega) \; \tilde{\in} \; \overline{D}_x] \geq v [\overline{x}(\omega) \; \tilde{\notin} \; D_x] = 1 - v [\overline{x}(\omega) \; \tilde{\in} \; D_x] \tag{1.22}$$

where $\overline{D}_x = X - D_x$.

As was said in Preface, the definition of uncertain logic should contain two parts: a mathematical model (which is described above) and its interpretation (semantics). The semantics is here the following: the uncertain logic operates with crisp predicates $P [\overline{x}(\omega)]$, but for the given ω it is not possible to state whether $P(\overline{x})$ is true or false because the function $\overline{x} = g(\omega)$ and consequently the value $\overline{x}$ corresponding to ω is unknown. The exact information, i.e. the knowledge of g is replaced by $h_\omega(x)$ which for the given ω characterizes the different possible approximate values of $\overline{x}(\omega)$. If we use the terms: knowledge, information, data etc., it is necessary to determine the subject (who knows?, who gives the information?).

In our considerations this subject is called *an expert*. So the expert does not know exactly the value $\bar{x}(\omega)$, but "looking at" ω he obtains some information concerning $\bar{x}$, which he does not express in an explicit form but uses it to formulate $h_\omega(x)$. Hence, the expert is the source of $h_\omega(x)$ which for particular x evaluates his opinion that $\bar{x} \cong x$. That is why $h_\omega(x)$ and consequently $v[\overline{\Psi}(\omega, P)]$ are called degrees of certainty. For example Ω is a set of persons, $\bar{x}(\omega)$ denotes the age of ω and the expert looking at the person ω gives the function $h_\omega(x)$ whose value for the particular x is his degree of certainty that the age of this person is approximately equal to x. The predicates $\overline{\Psi}(\omega, P)$ are soft because of the uncertainty of the expert. The result of including $h_\omega(x)$ into the definition of uncertain logic is that for the same (Ω, X) we may have the different logics specified by different experts.

The logic introduced by Definition 1.1 will be denoted by L-logic. In the next part we shall consider other versions of uncertain logic which will be denoted by L_p, L_n and L_c.

1.2 Other Versions of Uncertain Logic

Definition 1.2 (L_p-logic): The first part is the same as in Definition 1.1. The certainty index

$$v_p[\overline{\Psi}(\omega, P)] = v[\overline{\Psi}(\omega, P)] = \max_{x \in D_x} h_\omega(x).$$
(1.23)

The operations are defined in the following way

$$\neg \overline{\Psi}(\omega, P) = \overline{\Psi}(\omega, \neg P),$$
(1.24)

$$\overline{\Psi}(\omega, P_1) \vee \overline{\Psi}(\omega, P_2) = \overline{\Psi}(\omega, P_1 \vee P_2),$$
(1.25)

$$\overline{\Psi}(\omega, P_1) \wedge \overline{\Psi}(\omega, P_2) = \overline{\Psi}(\omega, P_1 \wedge P_2).$$
(1.26)

□

Consequently, we have the same equalities for v_p, i.e.

$$v_p[\neg \overline{\Psi}(\omega, P)] = v_p[\overline{\Psi}(\omega, \neg P)],$$
(1.27)

$$v_p[\overline{\Psi}(\omega, P_1) \vee \overline{\Psi}(\omega, P_2)] = v_p[\overline{\Psi}(\omega, P_1 \vee P_2)],$$
(1.28)

$$v_p[\overline{\Psi}(\omega, P_1) \wedge \overline{\Psi}(\omega, P_2)] = v_p[\overline{\Psi}(\omega, P_1 \wedge P_2)].$$
(1.29)

In a similar way as for L-logic it is easy to prove that:

$$\text{If} \quad P_1 \to P_2 \quad \text{then} \quad v_p[\overline{\Psi}(\omega, P_1)] \le v_p[\overline{\Psi}(\omega, P_2)], \tag{1.30}$$

$$v_p[\overline{\Psi}(\omega, P_1 \vee P_2)] = \max\{v_p[\overline{\Psi}(\omega, P_1)], v_p[\overline{\Psi}(\omega, P_2)]\}, \tag{1.31}$$

$$v_p[\overline{\Psi}(\omega, P_1 \wedge P_2)] \le \min\{v_p[\overline{\Psi}(\omega, P_1)], v_p[\overline{\Psi}(\omega, P_2)]\}, \tag{1.32}$$

$$v_p[\overline{\Psi}(\omega, \neg P)] \ge 1 - v_p[\overline{\Psi}(\omega, P)]. \tag{1.33}$$

Definition 1.3 (L_n-logic):   The certainty index of $\overline{\Psi}$ is defined as follows

$$v_n[\overline{\Psi}(\omega, P)] = 1 - v_p[\overline{\Psi}(\omega, \neg P)] = 1 - \max_{x \in \overline{D}_x} h_\omega(x). \tag{1.34}$$

The operations are the same as for v_p in L_p-logic, i.e. (1.24), (1.25), (1.26) and (1.27), (1.28), (1.29) with v_n in the place of v_p. □
It may be proved that:

$$\text{If} \quad P_1 \to P_2 \quad \text{then} \quad v_n[\overline{\Psi}(\omega, P_1)] \le v_n[\overline{\Psi}(\omega, P_2)], \tag{1.35}$$

$$v_n[\overline{\Psi}(\omega, P_1 \vee P_2)] \ge \max\{v_n[\overline{\Psi}(\omega, P_1)], v_n[\overline{\Psi}(\omega, P_2)]\}, \tag{1.36}$$

$$v_n[\overline{\Psi}(\omega, P_1 \wedge P_2)] = \min\{v_n[\overline{\Psi}(\omega, P_1)], v_n[\overline{\Psi}(\omega, P_2)]\} \quad \text{for} \quad w(P_1 \wedge P_2) > 0, \tag{1.37}$$

$$v_n[\overline{\Psi}(\omega, \neg P)] \le 1 - v_n[\overline{\Psi}(\omega, P)]. \tag{1.38}$$

The statement (1.35) follows immediately from (1.30) and (1.34). Property (1.36) follows from $D_1 \cup D_2 \supseteq D_1$, $D_1 \cup D_2 \supseteq D_2$ and (1.35). From (1.34) we have

$$v_n[\overline{\Psi}(\omega, P_1 \wedge P_2)] = 1 - \max_{x \in \overline{D}_1 \cup \overline{D}_2} h_\omega(x) = 1 - \max\{\max_{x \in \overline{D}_1} h_\omega(x), \max_{x \in \overline{D}_2} h_\omega(x)\}$$
$$= 1 - \max\{1 - v_n[\overline{\Psi}(\omega, P_1)], 1 - v_n[\overline{\Psi}(\omega, P_2)]\}$$

which proves (1.37). Substituting (1.34) into (1.33) we obtain (1.38).
In Definition 1.2 the certainty index is defined in "a positive way", so we may use the term: "positive" logic (L_p). In Definition 1.3 the certainty index is defined in "a negative way" and consequently we may use the term: "negative" logic (L_n). In (1.23) the shape of the function $h_\omega(x)$ in $\overline{D}_x$ is not taken into account, and in (1.34) the function $h_\omega(x)$ in D_x is neglected. They are the known disadvantages of these definitions. To avoid them consider the *combined logic* denoted by L_c.

Definition 1.4 (L_c-logic): The certainty index of $\overline{\varPsi}$ and the negation $\neg \overline{\varPsi}$ are defined as follows:

$$v_c[\overline{\varPsi}(\omega,P)] = \frac{v_p[\overline{\varPsi}(\omega,P)] + v_n[\overline{\varPsi}(\omega,P)]}{2} = \frac{1}{2}[\max_{x \in D_x} h_\omega(x) + 1 - \max_{x \in \overline{D}_x} h_\omega(x)],$$

(1.39)

$$\neg\overline{\varPsi}(\omega,P) = \overline{\varPsi}(\omega,\neg P).$$

(1.40)

The operations for v_c are determined by the operations for v_p and v_n.   □

Since L_c-logic will be used in the next section for the formulation of the uncertain variable, it will be described in more detail than L_p and L_n. According to (1.40)

$$v_c[\neg\overline{\varPsi}(\omega,P)] = v_c[\overline{\varPsi}(\omega,\neg P)].$$

Using (1.39) and (1.28), (1.29) for v_p and v_n, it is easy to show that

$$v_c[\overline{\varPsi}(\omega,P_1) \vee \overline{\varPsi}(\omega,P_2)] = v_c[\overline{\varPsi}(\omega,P_1 \vee P_2)],$$

(1.41)

$$v_c[\overline{\varPsi}(\omega,P_1) \wedge \overline{\varPsi}(\omega,P_2)] = v_c[\overline{\varPsi}(\omega,P_1 \wedge P_2)].$$

(1.42)

L_c-logic may be defined independently of v_p and v_n, with the right hand side of (1.39) and the definitions of operations (1.40), (1.41), (1.42). The operations may be rewritten in the following form

$$\overline{x} \tilde{\notin} D_x = \overline{x} \tilde{\in} \overline{D}_x,$$

(1.43)

$$v_c[\overline{x}(\omega) \tilde{\in} D_1 \vee \overline{x}(\omega) \tilde{\in} D_2] = v_c[\overline{x}(\omega) \tilde{\in} D_1 \cup D_2],$$

(1.44)

$$v_c[\overline{x}(\omega) \tilde{\in} D_1 \wedge \overline{x}(\omega) \tilde{\in} D_2] = v_c[\overline{x}(\omega) \tilde{\in} D_1 \cap D_2].$$

(1.45)

From (1.8), (1.34) and (1.39)

$$v_c[\overline{x}(\omega) \tilde{\in} X] = 1, \qquad v_c[\overline{x}(\omega) \tilde{\in} \varnothing] = 0.$$

(1.46)

One can note that $G_\omega(x) = "\overline{x} \cong x"$ is a special case of $\overline{\varPsi}$ for $D_x = \{x\}$ and according to (1.39)

$$v_c[\overline{x}(\omega) \cong x] = \frac{1}{2}[h_\omega(x) + 1 - \max_{\tilde{x} \in X - \{x\}} h_\omega(\tilde{x})],$$

(1.47)

$$v_c[\overline{x}(\omega) \ncong x] = \frac{1}{2}[\max_{\tilde{x} \in X - \{x\}} h_\omega(\tilde{x}) + 1 - h_\omega(x)].$$

(1.48)

It is worth noting that if $h_\omega(x)$ is a continuous function then

$$v_c[\bar{x}(\omega) \tilde{=} x] = \frac{1}{2}h_\omega(x) .$$

Using (1.30) and (1.35), we obtain the following property: If for each x $P_1 \rightarrow P_2$ (i.e. $D_1 \subseteq D_2$) then

$$v_c[\overline{\Psi}(\omega, P_1)] \leq v_c[\overline{\Psi}(\omega, P_2)] \quad \text{or} \quad v_c[\bar{x}(\omega) \tilde{\in} D_1] \leq v_c[\bar{x}(\omega) \tilde{\in} D_2] .(1.49)$$

Theorem 1.3:

$$v_c[\overline{\Psi}(\omega, P_1 \vee P_2)] \geq \max \{v_c[\overline{\Psi}(\omega, P_1)], \, v_c[\overline{\Psi}(\omega, P_2)]\}, \qquad (1.50)$$

$$v_c[\overline{\Psi}(\omega, P_1 \wedge P_2)] \leq \min \{v_c[\overline{\Psi}(\omega, P_1)], \, v_c[\overline{\Psi}(\omega, P_2)]\}. \qquad (1.51)$$

Proof: Inequality (1.50) may be obtained from $D_1 \cup D_2 \supseteq D_1$, $D_1 \cup D_2 \supseteq D_2$ and (1.49). Inequality (1.51) follows from $D_1 \cap D_2 \subseteq D_1$, $D_1 \cap D_2 \subseteq D_2$ and (1.49). The property (1.50) can also be delivered from (1.39), (1.31), (1.36), and the property (1.51) – from (1.39), (1.32), (1.37).   □

Theorem 1.4:

$$v_c[\neg \overline{\Psi}(\omega, P)] = 1 - v_c[\overline{\Psi}(\omega, P)]. \qquad (1.52)$$

Proof: From (1.34) and (1.39)

$$v_c[\overline{\Psi}(\omega, P)] = \frac{1}{2}\{v_p[\overline{\Psi}(\omega, P)] + 1 - v_p[\overline{\Psi}(\omega, \neg P)]\} .$$

Then

$$v_c[\neg \overline{\Psi}(\omega, P)] = \frac{1}{2}\{v_p[\overline{\Psi}(\omega, \neg P)] + 1 - v_p[\overline{\Psi}(\omega, P)]\} = 1 - v_c[\overline{\Psi}(\omega, P)]. \quad □$$

Till now it has been assumed that $\bar{x}(\omega), x \in X$. The considerations can be extended for the case $\bar{x}(\omega) \in X$ and $x \in X \subset \overline{X}$. It means that the set of approximate values X evaluated by an expert may be a subset of the set of the possible values of $\bar{x}(\omega)$. In a typical case $X = \{x_1, x_2, ..., x_m\}$ (a finite set), $x_i \in X$ for $i \in \overline{1,m}$. In our example with persons and age an expert may give the values $h_\omega(x)$ for natural numbers, e.g. $X = \{18, 19, 20, 21, 22\}$.

1.3 Uncertain Variables

The variable $\bar{x}$ for a fixed ω will be called an uncertain variable. Two versions of uncertain variables will be defined. The precise definition will contain: $h(x)$ given by an expert, the definition of the certainty index $w(\bar{x} \,\tilde{\in}\, D_x)$ and the definitions of $w(\bar{x} \,\tilde{\notin}\, D_x)$, $w(\bar{x} \,\tilde{\in}\, D_1 \vee \bar{x} \,\tilde{\in}\, D_2)$, $w(\bar{x} \,\tilde{\in}\, D_1 \wedge \bar{x} \,\tilde{\in}\, D_2)$.

Definition 1.5 (uncertain variable): The uncertain variable $\bar{x}$ is defined by the set of values X, the function $h(x) = v(\bar{x} \cong x)$ (i.e. the certainty index that $\bar{x} \cong x$, given by an expert) and the following definitions:

$$v(\bar{x} \,\tilde{\in}\, D_x) = \begin{cases} \max\limits_{x \in D_x} h(x) & \text{for } D_x \neq \varnothing \\ 0 & \text{for } D_x = \varnothing, \end{cases} \tag{1.53}$$

$$v(\bar{x} \,\tilde{\notin}\, D_x) = 1 - v(\bar{x} \,\tilde{\in}\, D_x), \tag{1.54}$$

$$v(\bar{x} \,\tilde{\in}\, D_1 \vee \bar{x} \,\tilde{\in}\, D_2) = \max\{v(\bar{x} \,\tilde{\in}\, D_1), v(\bar{x} \,\tilde{\in}\, D_2)\}, \tag{1.55}$$

$$v(\bar{x} \,\tilde{\in}\, D_1 \wedge \bar{x} \,\tilde{\in}\, D_2) = \begin{cases} \min\{v(\bar{x} \,\tilde{\in}\, D_1), v(\bar{x} \,\tilde{\in}\, D_2)\} & \text{for } D_1 \cap D_2 \neq \varnothing \\ 0 & \text{for } D_1 \cap D_2 = \varnothing. \end{cases} \tag{1.56}$$

The function $h(x)$ will be called a *certainty distribution*. □

The definition of the uncertain variable is based on the uncertain logic, i.e. L-logic (see Definition 1.1). Then the properties (1.17), (1.18), (1.19), (1.20), (1.22) are satisfied. The properties (1.19) and (1.20) may be presented in the following form

$$v(\bar{x} \,\tilde{\in}\, D_1 \cup D_2) = \max\{v(\bar{x} \,\tilde{\in}\, D_1), v(\bar{x} \,\tilde{\in}\, D_2)\}, \tag{1.57}$$

$$v(\bar{x} \,\tilde{\in}\, D_1 \cap D_2) \leq \min\{v(\bar{x} \,\tilde{\in}\, D_1), v(\bar{x} \,\tilde{\in}\, D_2)\}. \tag{1.58}$$

Example 1.1: $X = \{1, 2, 3, 4, 5, 6, 7\}$ and the corresponding values of $h(x)$ are $(0.5, 0.8, 1, 0.6, 0.5, 0.4, 0.2)$, i.e. $h(1) = 0.5$, $h(2) = 0.8$ etc. Let $D_1 = \{1, 2, 4, 5, 6\}$, $D_2 = \{3, 4, 5\}$. Then $D_1 \cup D_2 = \{1, 2, 3, 4, 5, 6\}$, $D_1 \cap D_2 = \{4, 5\}$, $v(\bar{x} \,\tilde{\in}\, D_1) = \max\{0.5, 0.8, 0.6, 0.5, 0.4\} = 0.8$, $v(\bar{x} \,\tilde{\in}\, D_2) = \max\{1, 0.6, 0.5\} = 1$, $v(\bar{x} \,\tilde{\in}\, D_1 \cup D_2) = \max\{0.5, 0.8, 1, 0.6, 0.5, 0.4\} = 1$ $v(\bar{x} \,\tilde{\in}\, D_1 \vee \bar{x} \,\tilde{\in}\, D_2) = \max\{0.8, 1\} = 1$, $v(\bar{x} \,\tilde{\in}\, D_1 \cap D_2) = \max\{0.6, 0.5\} = 0.6$, $v(\bar{x} \,\tilde{\in}\, D_1 \wedge \bar{x} \,\tilde{\in}\, D_2) = \min\{0.8, 1\} = 0.8$.

Example 1.2:   The certainty distribution is shown in Fig. 1.1. Let $D_x = [0, 4]$.
Then

$$v(\bar{x} \,\tilde{\in}\, D_x) = v(\bar{x} \,\tilde{\in}\, [0,4]) = 0.8,$$

$$v(\bar{x} \,\tilde{\in}\, \overline{D}_x) = v(\bar{x} \,\tilde{\in}\, [4,16]) = 1,$$

$$v(\bar{x} \,\tilde{\notin}\, \overline{D}_x) = v(\bar{x} \,\tilde{\notin}\, [4,16]) = 1-1 = 0 < v(\bar{x} \,\tilde{\in}\, D_x),$$

$$v(\bar{x} \,\tilde{\notin}\, D_x) = v(\bar{x} \,\tilde{\notin}\, [0,4]) = 1-0.8 = 0.2 < v(\bar{x} \,\tilde{\in}\, \overline{D}_x).$$

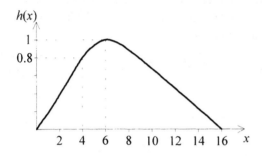

Fig. 1.1. Example of certainty distribution

Definition 1.6 (C-uncertain variable): C-uncertain variable $\bar{x}$ is defined by the
set of values X, the function $h(x) = v(\bar{x} \cong x)$ given by an expert, and the following definitions:

$$v_c(\bar{x} \,\tilde{\in}\, D_x) = \frac{1}{2}[\max_{x \in D_x} h(x) + 1 - \max_{x \in \overline{D}_x} h(x)], \tag{1.59}$$

$$v_c(\bar{x} \,\tilde{\notin}\, D_x) = 1 - v_c(\bar{x} \,\tilde{\in}\, D_x), \tag{1.60}$$

$$v_c(\bar{x} \,\tilde{\in}\, D_1 \vee \bar{x} \,\tilde{\in}\, D_2) = v_c(\bar{x} \,\tilde{\in}\, D_1 \cup D_2), \tag{1.61}$$

$$v_c(\bar{x} \,\tilde{\in}\, D_1 \wedge \bar{x} \,\tilde{\in}\, D_2) = v_c(\bar{x} \,\tilde{\in}\, D_1 \cap D_2). \tag{1.62}$$

$\square$

The definition of C-uncertain variable is based on L_c-logic (see Definition 1.4).
Then the properties (1.46), (1.47), (1.48) are satisfied. According to (1.40) and
(1.52)

$$v_c(\bar{x} \,\tilde{\notin}\, D_x) = v_c(\bar{x} \,\tilde{\in}\, \overline{D}_x). \tag{1.63}$$

Inequalities (1.50) and (1.51) may be presented in the following form

$$v_c(\bar{x} \,\tilde{\in}\, D_1 \cup D_2) \geq \max\{v_c(\bar{x} \,\tilde{\in}\, D_1), v_c(\bar{x} \,\tilde{\in}\, D_2)\}, \tag{1.64}$$

$$v_c(\bar{x} \,\tilde{\in}\, D_1 \cap D_2) \leq \min\{v_c(\bar{x} \,\tilde{\in}\, D_1), v_c(\bar{x} \,\tilde{\in}\, D_2)\}. \tag{1.65}$$

The function $v_c(\overline{x} \cong x) \triangleq h_c(x)$ expressed by (1.47) may be called a C-certainty distribution. Note that the certainty distribution $h(x)$ is given by an expert and C-certainty distribution may be determined according to (1.47), using $h(x)$. The C-certainty distribution does not determine the certainty index $v_c(\overline{x} \tilde{\in} D_x)$. To determine v_c, it is necessary to know $h(x)$ and to use (1.59). According to (1.64)

$$\max_{x \in D_x} h_c(x) \le v_c(\overline{x} \tilde{\in} D_x) .$$

The formula (1.59) may be presented in the following way

$$v_c(\overline{x} \tilde{\in} D_x) = \begin{cases} \dfrac{1}{2} \max_{x \in D_x} h(x) = \dfrac{1}{2} v(\overline{x} \tilde{\in} D_x) & \text{if} \quad \max_{x \in \overline{D}_x} h(x) = 1 \\ 1 - \dfrac{1}{2} \max_{x \in \overline{D}_x} h(x) = v(\overline{x} \tilde{\in} D_x) - \dfrac{1}{2} v(\overline{x} \tilde{\in} \overline{D}_x) & \text{otherwise.} \end{cases}$$

(1.66)

The formula (1.66) shows the relation between the certainty indexes v and v_c for the same D_x: if $D_x \ne X$ and $D_x \ne \varnothing$ then $v_c < v$. In particular, (1.47) becomes

$$h_c(x) = \begin{cases} \dfrac{1}{2} h(x) & \text{if} \quad \max_{\overline{x} \in X - \{x\}} h(\overline{x}) = 1 \\ 1 - \dfrac{1}{2} \max_{\overline{x} \in X - \{x\}} h(\overline{x}) & \text{otherwise.} \end{cases}$$

(1.67)

In the continuous case

$$h_c(x) = \frac{1}{2} h(x)$$

and in the discrete case

$$h_c(x_i) = \begin{cases} \dfrac{1}{2} h(x_i) & \text{if} \quad \max_{x \ne x_i} h(x) = 1 \\ 1 - \dfrac{1}{2} \max_{x \ne x_i} h(x) & \text{otherwise.} \end{cases}$$

Example 1.3: The set X and $h(x)$ are the same as in Example 1.1. Using (1.67) we obtain $h_c(1) = 0.25$, $h_c(2) = 0.4$, $h_c(3) = 1 - \dfrac{0.8}{2} = 0.6$, $h_c(4) = 0.3$, $h_c(5) = 0.25$, $h_c(6) = 0.2$, $h_c(7) = 0.1$. Let D_1 and D_2 be the same as in Example 1.1. Using (1.66) and the values v obtained in Example 1.1 we have:

$$v_c(\overline{x} \tilde{\in} D_1) = \frac{1}{2} v = 0.4, \quad v_c(\overline{x} \tilde{\in} D_2) = 1 - \frac{0.8}{2} = 0.6, \quad v_c(\overline{x} \tilde{\in} D_1 \vee \overline{x} \tilde{\in} D_2) =$$

$$v_c(\overline{x} \,\widetilde{\in}\, D_1 \cup D_2) = 1 - \frac{0.2}{2} = 0.9 \;,$$

$v_c(\overline{x} \,\widetilde{\in}\, D_1 \wedge \overline{x} \,\widetilde{\in}\, D_2) = v_c(\overline{x} \,\widetilde{\in}\, D_1 \cap D_2) = \dfrac{0.6}{2} = 0.3$. In this case, for both D_1 and D_2, $v_c(\overline{x} \,\widetilde{\in}\, D) = \max h_c(x)$ for $x \in D$. Let $D = \{2, 3, 4\}$. Now $v_c(\overline{x} \,\widetilde{\in}\, D) = 1 - \dfrac{0.5}{2} = 0.75$ and $\max h_c(x) = \max \{0.4,\, 0.6,\, 0.3\} = 0.6 < v_c$.

Example 1.4:   The certainty distribution and D_x are the same as in Example 1.2.

$$v_c(\overline{x} \,\widetilde{\in}\, D_x) = \frac{1}{2}[v(\overline{x} \,\widetilde{\in}\, D_x) + v(\overline{x} \,\widetilde{\notin}\, \overline{D}_x)] = \frac{1}{2}[0.8 + 0] = 0.4\;,$$

$$v_c(\overline{x} \,\widetilde{\in}\, \overline{D}_x) = v_c(\overline{x} \,\widetilde{\notin}\, D_x) = 1 - v_c(\overline{x} \,\widetilde{\in}\, D_x) = 0.6\;,$$

$$v_c(\overline{x} \,\widetilde{\notin}\, \overline{D}_x) = v_c(\overline{x} \,\widetilde{\in}\, D_x) = 0.4,$$

$$v(\overline{x} \,\widetilde{\notin}\, D_x) = v_c(\overline{x} \,\widetilde{\in}\, \overline{D}_x) = 0.6.$$

The uncertain logic and L_c-logic have been chosen as the bases for the uncertain variable and C-uncertain variable, respectively, because of the advantages of these approaches. In both cases the logic value of the negation is $w(\overline{x} \,\widetilde{\notin}\, D_x) = 1 - w(\overline{x} \,\widetilde{\in}\, D_x)$ (see (1.54) and (1.60)). In the first case it is easy to determine the certainty indexes for $\overline{x} \,\widetilde{\in}\, D_1 \vee \overline{x} \,\widetilde{\in}\, D_2$ and $\overline{x} \,\widetilde{\in}\, D_1 \wedge \overline{x} \,\widetilde{\in}\, D_2$ and all operations are the same as in (1.1) for multi-valued logic. In the second case in the definition of the certainty index $v_c(\overline{x} \,\widetilde{\in}\, D_x)$ the values of $h(x)$ for $\overline{D}_x$ are also taken into account and the logic operations (negation, disjunction and conjunction) correspond to the operations in the family of subsets D_x (complement, union and intersection). On the other hand, the calculations of the certainty indexes for disjunction and conjunction are more complicated than in the first case and are not determined by the certainty indexes for $\overline{x} \,\widetilde{\in}\, D_1$, $\overline{x} \,\widetilde{\in}\, D_2$, i.e. they cannot be reduced to operations in the set of certainty indexes $v_c(\overline{x} \,\widetilde{\in}\, D)$. These features should be taken into account when making a choice between the application of the uncertain variable or C-uncertain variable in particular cases.

1.4 Additional Description of Uncertain Variables

For the further considerations we assume $X \subseteq R^k$ (k-dimensional real number vector space) and we shall consider two cases: the discrete case with $X = \{x_1, x_2, ..., x_m\}$ and the continuous case in which $h(x)$ is a continuous function.

Definition 1.7: In the discrete case

$$\bar{h}(x_i) = \frac{h(x_i)}{\sum\limits_{j=1}^{m} h(x_j)}, \qquad i \in \overline{1,m} \tag{1.68}$$

will be called a *normalized certainty distribution*. The value

$$M(\bar{x}) = \sum_{i=1}^{m} x_i \, \bar{h}(x_i) \tag{1.69}$$

will be called a *mean value* of the uncertain variable $\bar{x}$. In the continuous case the normalized certainty distribution and the mean value are defined as follows:

$$\bar{h}(x) = \frac{h(x)}{\int\limits_X h(x)\,dx}, \qquad M(\bar{x}) = \int_X x\,\bar{h}(x)\,dx . \tag{1.70}$$

For C-uncertain variable the normalized C-certainty distribution $\bar{h}_c(x)$ and the mean value $M_c(\bar{x})$ are defined in the same way, with h_c in the place of h in (1.68), (1.69) and (1.70). $\qquad\qquad\qquad\qquad\qquad\qquad\qquad\qquad\qquad$ □

In the continuous case $h_c(x) = \frac{1}{2}h(x)$, then $\bar{h}_c(x) = \bar{h}(x)$ and $M_c = M$. In the discrete case, if x^* is a unique value for which $h(x^*) = 1$ and

$$\max_{x \neq x^*} h(x) \approx 1$$

then $M_c \approx M$. As a value characterizing $h(x)$ or $h_c(x)$ one can also use

$$x^* = \arg\max_{x \in X} h(x) \qquad \text{or} \qquad x_c^* = \arg\max_{x \in X} h_c(x) .$$

Replacing the uncertain variable $\bar{x}$ by its deterministic representation $M(\bar{x})$ or x^* may be called a *determinization* (analogous to defuzzification for fuzzy numbers).

Consider now a function $\Phi : X \to Y$, $Y \subseteq R^k$, i.e. $y = \Phi(x)$. We say that the uncertain variable $\bar{y} = < Y, h_y(y) >$ is a function of the uncertain variable $\bar{x} = < X, h_x(x) >$, i.e. $\bar{y} = \Phi(\bar{x})$ where the certainty distribution $h_y(y)$ is determined by $h_x(x)$ and Φ:

$$h_y(y) = v(\bar{y} \cong y) = \max_{x \in D_x(y)} h_x(x) \tag{1.71}$$

where

$$D_x(y) = \{x \in X : \Phi(x) = y\} .$$

If $y = \Phi(x)$ is one-to-one mapping and $x = \Phi^{-1}(y)$ then

$$D_x(y) = \{ \Phi^{-1}(y) \}$$

and

$$h_y(y) = h_x[\Phi^{-1}(y)] .$$

In this case, according to (1.68) and (1.69)

$$M_y(\bar{y}) = \sum_{i=1}^{m} \Phi(x_i) h_x(x_i) [\sum_{j=1}^{m} h_x(x_j)]^{-1} .$$

For C-uncertain variables C-certainty distribution $h_{cy}(y) = v_c(\bar{y} \cong y)$ may be determined in two ways:

1. According to (1.67)

$$h_{cy}^{I}(y) = \begin{cases} \dfrac{1}{2} h_y(y) & \text{if } \max_{\bar{y} \in Y - \{y\}} h_y(\bar{y}) = 1 \\ 1 - \dfrac{1}{2} \max_{\bar{y} \in Y - \{y\}} h_y(\bar{y}) & \text{otherwise} . \end{cases} \tag{1.72}$$

where $h_y(y)$ is determined by (1.71).

2. According to (1.66)

$$h_{cy}^{II}(y) = \begin{cases} \dfrac{1}{2} \max_{x \in D_x(y)} h_x(x) & \text{if } \max_{x \in \bar{D}_x} h_x(x) = 1 \\ 1 - \dfrac{1}{2} \max_{x \in \bar{D}_x(y)} h_x(x) & \text{otherwise} . \end{cases} \tag{1.73}$$

Theorem 1.5:

$$h_{cy}^{I}(y) = h_{cy}^{II}(y) .$$

Proof: It is sufficient to prove that

$$\max_{\bar{y} \in Y - \{y\}} h_y(\bar{y}) = \max_{x \in \bar{D}_x} h_x(x) .$$

From (1.71)

$$\max_{\bar{y} \in Y - \{y\}} h_y(\bar{y}) = \max_{\bar{y} \in Y - \{y\}} [\max_{x \in D_x(\bar{y})} h_x(x)] .$$

Note that if $y_1 \neq y_2$ then $D_x(y_1) \cap D_x(y_2) = \emptyset$. Consequently, $D_x(\bar{y}) \cap D_x(y) = \emptyset$ and

$$\bigcup_{\bar{y} \in Y - \{y\}} D_x(\bar{y}) = \overline{D}_x(y) .$$

Therefore

$$\max_{\bar{y} \in Y - \{y\}} [\max_{x \in D_x(\bar{y})} h_x(x)] = \max_{x \in \overline{D}_x(y)} h_x(x) . \qquad \square$$

It is important to note that $h_{cy}(y)$ is not determined by $h_{cx}(x)$. To determine $h_{cy}(y)$ it is necessary to know $h_x(x)$ and to use (1.73), or (1.71) and (1.72).

Let us now consider a pair of uncertain variables $(\bar{x}, \bar{y}) = < X \times Y, h(x, y) >$ where $h(x, y) = v[(\bar{x}, \bar{y}) \cong (x, y)]$ is given by an expert and is called a *joint certainty distribution*. Then, using (1.1) for the disjunction in multi-valued logic, we have the following *marginal certainty distributions*

$$h_x(x) = v(\bar{x} \cong x) = \max_{y \in Y} h(x, y), \qquad (1.74)$$

$$h_y(y) = v(\bar{y} \cong y) = \max_{x \in X} h(x, y) . \qquad (1.75)$$

If the certainty index $v[\bar{x}(\omega) \cong x]$ given by an expert depends on the value of y for the same ω (i.e. if the expert changes the value $h_x(x)$ when he obtains the value y for the element ω "under observation") then $h_x(x|y)$ may be called a *conditional certainty distribution*. The variables $\bar{x}, \bar{y}$ are called independent when

$$h_x(x|y) = h_x(x), \qquad h_y(y|x) = h_y(y) .$$

Using (1.1) for the conjunction in multi-valued logic we obtain

$$h(x, y) = v(\bar{x} \cong x \wedge \bar{y} \cong y) = \min\{h_x(x), h_y(y|x)\} = \min\{h_y(y), h_x(x|y)\}. \qquad (1.76)$$

Taking into account the relationships between the certainty distributions one can see that they cannot be given independently by an expert. If the expert gives $h_x(x)$ and $h_y(y|x)$ or $h_y(y)$ and $h_x(x|y)$ then $h(x, y)$ is already determined by (1.76). The joint distribution $h(x, y)$ given by an expert determines $h_x(x)$ (1.74) and $h_y(y)$ (1.75) but does not determine $h_x(x|y)$ and $h_y(y|x)$. In such a case only sets of functions $h_x(x|y)$ and $h_y(y|x)$ satisfying (1.76) are deter-

mined. For the function $\bar{y} = \Phi(\bar{x})$ where $\bar{x}$ is a pair of variables $(\bar{x}_1, \bar{x}_2)$,
$x_{1,2} \in X$, according to (1.71)

$$h_y(y) = \max_{(x_1, x_2) \in D(y)} h(x_1, x_2) \ ,$$

$h(x_1, x_2)$ is determined by (1.76) for $x = x_1$, $y = x_2$, and

$$D(y) = \{(x_1, x_2) \in X \times X : \Phi(x_1, x_2) = y\} \ .$$

2 Analysis and Decision Making for Static Plants

2.1 Analysis Problem for a Functional Plant

Let us consider a static plant with input vector $u \in U$ and output vector $y \in Y$, where U and Y are real number vector spaces. When the plant is described by a function $y = \Phi(u)$, the analysis problem consists in finding the value y for the given value u. Consider now the plant described by $y = \Phi(u, x)$ where $x \in X$ is an unknown vector parameter which is assumed to be a value of an uncertain variable $\bar{x}$ with the certainty distribution $h_x(x)$ given by an expert. Then y is a value of an uncertain variable $\bar{y}$ and for the fixed u, $\bar{y}$ is the function of $\bar{x}$: $\bar{y} = \Phi(u, \bar{x})$.

Analysis problem may be formulated as follows: For the given Φ, $h_x(x)$ and u find the certainty distribution $h_y(y)$ of the uncertain variable $\bar{y}$. Having $h_y(y)$ one can determine M_y and

$$y^* = \arg \max_{y \in Y} h_y(y), \quad \text{i.e.} \quad h_y(y^*) = 1.$$

According to (1.71)

$$h_y(y; u) = v(\bar{y} \cong y) = \max_{x \in D_x(y; u)} h_x(x) \tag{2.1}$$

where $D_x(y; u) = \{x \in X : \Phi(u, x) = y\}$. If Φ as a function of x is one-to-one mapping and $x = \Phi^{-1}(u, y)$ then

$$h_y(y; u) = h_x[\Phi^{-1}(u, y)]$$

and $y^* = \Phi(u, x^*)$ where $x^* = \arg \max h_x(x)$. From the definition of the certainty distributions h and h_c it is easy to note that in both continuous and discrete cases

$y^* = y_c^*$ where $y_c^* = \arg \max h_{cy}(y)$ and $h_{cy}(y)$ is a certainty distribution of $\overline{y}$ considered as C-uncertain variable.

Example 2.1: Let $u, x \in R^2$, $u = (u^{(1)}, u^{(2)})$, $x = (x^{(1)}, x^{(2)})$, $y \in R^1$,

$$y = x^{(1)}u^{(1)} + x^{(2)}u^{(2)},$$

$x^{(1)} \in \{3, 4, 5, 6\}$, $x^{(2)} \in \{5, 6, 7\}$ and the corresponding values of h_{x1}, h_{x2} given by an expert are $(0.3, 0.5, 1, 0.6)$ for $x^{(1)}$ and $(0.8, 1, 0.4)$ for $x^{(2)}$. Assume that $\overline{x}^{(1)}$ and $\overline{x}^{(2)}$ are independent, i.e. $h_x(x_i^{(1)}, x_j^{(2)}) = \min\{h_{x1}(x_i^{(1)}), h_{x2}(x_j^{(2)})\}$. Then for $x = (x^{(1)}, x^{(2)}) \in \{(3,5), (3,6), (3,7), (4,5), (4,6), (4,7), (5,5), (5,6), (5,7), (6,5), (6,6), (6,7)\}$ the corresponding values of h_x are $(0.3, 0.3, 0.3, 0.5, 0.5, 0.4, 0.8, 1, 0.4, 0.6, 0.6, 0.4)$. Let $u^{(1)} = 2$, $u^{(2)} = 1$. The values of $y = 2x^{(1)} + x^{(2)}$ corresponding to the set of pairs $(x^{(1)}, x^{(2)})$ are the following: $\{11, 12, 13, 13, 14, 15, 15, 16, 17, 17, 18, 19\}$. Then $h_y(11) = h_x(3,5) = 0.3$, $h_y(12) = h_x(3,6) = 0.3$, $h_y(13) = \max \{h_x(3,7), h_x(4,5)\} = 0.5$, $h_y(14) = h_x(4,6) = 0.5$, $h_y(15) = \max \{h_x(4,7), h_x(5,5)\} = 0.8$, $h_y(16) = h_x(5,6) = 1$, $h_y(17) = \max \{h_x(5,7), h_x(6,5)\} = 0.6$, $h_y(18) = h_x(6,6) = 0.6$, $h_y(19) = h_x(6,7) = 0.4$. For $h_y(y)$ we have $y^* = 16$. Using (1.68) and (1.69) for $\overline{y}$ we obtain $\overline{h}_y = 5$,

$$M_y = \frac{77}{5} = 15.40.$$

Using (1.72) we obtain the corresponding values of $h_{yc}(y)$: $(0.15, 0.15, 0.25, 0.25, 0.4, 1 - \frac{0.8}{2} = 0.6, 0.3, 0.3, 0.2)$. Then $y_c^* = y^* = 16$, $v_c(\overline{y} \cong 16) = 0.6$, $\overline{h}_{yc} = 2.6$, $M_{yc} = 15.43 \approx M_y$. □

2.2   Decision Making Problem for a Functional Plant

For the functional system $y = \Phi(u)$ the basic decision problem consists in finding the decision $\hat{u}$ for the given desirable value $\hat{y}$. Consider now the system with the unknown parameter x, described in 2.1.

Decision problem may be formulated as follows:

Version I: To find the decision $\hat{u}$ maximizing $v(\overline{y} \cong \hat{y})$.

Version II: To find $\hat{u}$ such that $M_y(\bar{y};u) = \hat{y}$ where M_y is the mean value of

$\bar{y} = \Phi(u,\bar{x})$ for the fixed u.

Version III: To find $\hat{u}$ minimizing $M_s(\bar{s};u)$ where $s = \varphi(y,\hat{y})$ is a quality in-

dex, e.g. $s = (y-\hat{y})^T(y-\hat{y})$ where y and $\hat{y}$ are column vectors.

When $\bar{x}$ is assumed to be a C-uncertain variable, the formulations are the same with v_c, M_{cy}, M_{cs} instead of v, M_y, M_s. It is worth noting that the decision problem statements are analogous to those in the probabilistic approach where x is assumed to be a value of a random variable with the known probability distribution. In each version $h_y(y;u)$ should be determined according to (2.1). Then, in version I $\hat{u}$ is the value of u maximizing $h_y(\hat{y};u)$, i.e. $\hat{u}$ is the solution of the equation $\varepsilon(u) = \hat{y}$ where $y^* = \varepsilon(u)$ is a value of y maximizing $h_y(y;u)$. In version II $\hat{u}$ is obtained as a solution of the equation $M_y(\bar{y};u) = \hat{y}$.
In version III for the determination of $M_s(\bar{s};u)$ one should find

$$h_s(s;u) = \max_{y \in D_y(s)} h_y(y;u) \tag{2.2}$$

where $D_y(s) = \{y \in Y : \varphi(y,\hat{y}) = s\}$.

When $\bar{x}$ is considered as C-uncertain variable, it is necessary to determine h_{cy} using (1.72) or (1.73) and in version II to find $M_{cy}(\bar{y};u)$. In version III it is necessary to find h_{cs} according to (1.72) or (1.73) with (s,x) instead of (y,x) and then to determine $M_{cs}(\bar{s};u)$. Using (1.72) it is easy to see that if y^* is the only value for which $h_y = 1$ then $y_c^* = y^*$ where y_c^* is the value maximizing h_{cy}, and $M_{cy} \approx M_y$. Consequently, in this case the results $\hat{u}$ in version I are the same and in version II are approximately the same for the uncertain and C-uncertain variable, and for $u = \hat{u}$ in version I $v(\bar{y} \cong \hat{y}) = 1$, $v_c(\bar{y} \cong \hat{y}) < 1$.

Example 2.2: Let u, y, $x \in R^1$, $y = xu$, $X = \{3, 4, 5, 6, 7\}$ and the corresponding values of $h_x(x)$ are $(0.5, 1, 0.3, 0.1, 0.1)$. Using (1.67) or (1.72) for x we obtain the corresponding values of h_{cx}: $(0.25, 0.75, 0.15, 0.05, 0.05)$.
In version I $y^* = 4u$, i.e. $\hat{u} = 0.25\hat{y}$, $v(\bar{y} \cong \hat{y}) = 1$, $v_c(\bar{y} \cong \hat{y}) = 0.75$.

In our example $Y = \{3u, 4u, 5u, 6u, 7u\}$, the values of $h_y(y;u)$ are the same as h_x, the values of $\bar{h}_y$ are $(0.25, 0.5, 0.15, 0.05, 0.05)$ and the values of $h_{cy} = 1.25\bar{h}_{cy}$ are the same as h_{cx}. In version II, using (1.69) for y we obtain $M_y = 4.15u$ and $M_{cy} = 4.12u$. Then in both cases the result is approximately the same: $\hat{u} \approx 0.24\hat{y}$.

Let in version III $s = (y - \hat{y})^2$. Then $M_s(\bar{s};u)$ is equal to

$$0.25(3u - \hat{y})^2 + 0.5(4u - \hat{y})^2 + 0.15(5u - \hat{y})^2 + 0.05(6u - \hat{y})^2 + 0.05(7u - \hat{y})^2$$

$\overset{\triangle}{=} \overline{M}_s(u)$ except for

$$u = \frac{2\hat{y}}{x_i + x_j} \overset{\triangle}{=} u_d, \quad x_i \neq x_j, \quad x_i, x_j \in \{3, 4, 5, 6, 7\}. \quad (2.3)$$

For example, for $x_1 = 3$, $x_2 = 4$ and $u = u_d$ we have $(3u - \hat{y})^2 = (4u - \hat{y})^2$. Consequently, $s \in \{(4u - \hat{y})^2, (5u - \hat{y})^2, (6u - \hat{y})^2, (7u - \hat{y})^2\}$, according to (2.2) the values of h_s are $[\max(0.5, 1), 0.3, 0.1, 0.1]$ and

$$M_s(\bar{s};u) = \frac{1}{1.5}[(4u_d - \hat{y})^2 + 0.3(5u_d - \hat{y})^2 + 0.1(6u_d - \hat{y})^2 + 0.1(7u_d - \hat{y})^2]$$

Then $M_s(\bar{s};u)$ is a discontinuous function of u. The value of u minimizing $\overline{M}_s(u)$ is

$$u_{min} = \frac{4.15}{15.8}\hat{y} \approx 0.26\hat{y}.$$

From the sensitivity point of view it is reasonable not to take into account the points of discontinuity u_d (2.3) and to accept $\hat{u} = u_{min}$.

2.3   External Disturbances

The considerations may be easily extended for a plant with external disturbances, described by a function

$$y = \Phi(u, z, x) \quad (2.4)$$

where $z \in Z$ is a vector of the disturbances which can be measured.

Analysis problem: For the given Φ, $h_x(x)$, u and z find the certainty distribution $h_y(y)$.

According to (1.71)

$$h_y(y;u,z) = v(\bar{y} \cong y) = v[\bar{x} \tilde{\in} D_x(y;u,z)] = \max_{x \in D_x(y;u,z)} h_x(x) \qquad (2.5)$$

where

$$D_x(y;u,z) = \{x \in X : \Phi(u,z,x) = y\}.$$

Having $h_y(y;u,z)$ one can determine the mean value

$$M_y(\bar{y};u,z) = \int_Y y h_y(y;u,z)dy \cdot [\int_Y h_y(y;u,z)dy]^{-1} \triangleq \Phi_b(u,z) \qquad (2.6)$$

(for the continuous case) and

$$y^* = \arg \max_{y \in Y} h_y(y;u,z),$$

i.e. such a value y^* that $h_y(y^*;u,z) = 1$. If Φ as a function of x is one-to-one mapping and $x = \Phi^{-1}(u,z,y)$ then

$$h_y(y;u,z) = h_x[\Phi^{-1}(u,z,y)] \qquad (2.7)$$

and $y^* = \Phi(u,z,x^*)$ where x^* satisfies the equation $h_x(x) = 1$. It is easy to note that $y^* = y_c^*$ where

$$y_c^* = \arg \max_{y \in Y} h_{cy}(y;u,z)$$

and h_{cy} is a certainty distribution for C-uncertain variable.

Decision problem: For the given Φ, $h_x(x)$, z and $\hat{y}$

I. One should find $u \triangleq u_a$ maximizing $v(\bar{y} \cong \hat{y})$.

II. One should find $u \triangleq u_b$ such that $M_y(\bar{y}) = \hat{y}$.

The versions I, II correspond to the versions I, II of the decision problem presented in 2.2. The third version presented in 2.2 is much more complicated and will not be considered. In version I

$$u_a = \arg \max_{u \in U} \Phi_a(u, z) \overset{\Delta}{=} \Psi_a(z) \tag{2.8}$$

where $\Phi_a(u, z) = h_y(\hat{y}; u, z)$ and h_y is determined according to (2.5). The result u_a is a function of z if u_a is a unique value maximizing Φ_a for the given z.

In version II one should solve the equation

$$\Phi_b(u, z) = \hat{y} \tag{2.9}$$

where the function Φ_b is determined by (2.6). If the equation (2.9) has a unique solution with respect to u for a given z then as a result one obtains $u_b = \Psi_b(z)$. The functions Ψ_a and Ψ_b are two versions of the decision algorithm $u = \Psi(z)$ in an open-loop decision system (Fig. 2.1). It is worth noting that u_a is a decision for which $v(\overline{y} \cong \hat{y}) = 1$.

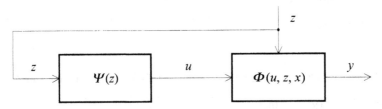

Fig. 2.1. Open-loop decision system

The functions Φ_a, Φ_b are results of two different ways of *determinization* of the uncertain plant, and the functions Ψ_a, Ψ_b are the respective decision algorithms based on the *knowledge of the plant* (KP):

$$KP = < \Phi, h_x >. \tag{2.10}$$

Assume that the equation

$$\Phi(u, z, x) = \hat{y}$$

has a unique solution with respect to u:

$$u \overset{\Delta}{=} \Phi_d(z, x). \tag{2.11}$$

The relationship (2.11) together with the certainty distribution $h_x(x)$ may be considered as a *knowledge of the decision making* (KD):

$$KD = < \Phi_d, h_x >, \tag{2.12}$$

obtained by using KP and $\hat{y}$. The equation (2.11) together with h_x may also be called an *uncertain decision algorithm* in the open-loop decision system. The determinization of this algorithm leads to two versions of the deterministic decision algorithm Ψ_d, corresponding to versions I and II of the decision problem:

I.

$$u_{ad} = \arg\max_{u \in U} h_u(u;z) \overset{\Delta}{=} \Psi_{ad}(z) \tag{2.13}$$

where

$$h_u(u;z) = \max_{x \in D_x(u;z)} h_x(x) \tag{2.14}$$

and

$$D_x(u;z) = \{x \in X : u = \Phi_d(z,x)\}. $$

II.

$$u_{bd} = M_u(\bar{u};z) \overset{\Delta}{=} \Psi_{bd}(z). \tag{2.15}$$

The decision algorithms Ψ_{ad} and Ψ_{bd} are based directly on the knowledge of the decision making. Two concepts of the determination of deterministic decision algorithms are illustrated in Figs. 2.2 and 2.3. In the first case (Fig. 2.2) the decision algorithms $\Psi_a(z)$ and $\Psi_b(z)$ are obtained via the determinization of the knowledge of the plant KP. In the second case (Fig. 2.3) the decision algorithms $\Psi_{ad}(z)$ and $\Psi_{bd}(z)$ are based on the determinization of the knowledge of the decision making KD obtained from KP for the given $\hat{y}$. The results of these two approaches may be different.

Theorem 2.1: For the plant described by KP in the form (2.10) and for KD in the form (2.12), if there exists an inverse function $x = \Phi^{-1}(u,z,y)$ then

$$\Psi_a(z) = \Psi_{ad}(z).$$

Proof: According to (2.7) and (2.13)

$$h_y(\hat{y};u,z) = h_x[\Phi^{-1}(u,z,\hat{y})],$$

$$h_u(u;z) = h_x[\Phi^{-1}(u,z,\hat{y})].$$

Then, using (2.8) and (2.13) we obtain $\Psi_a(z) = \Psi_{ad}(z)$.   □

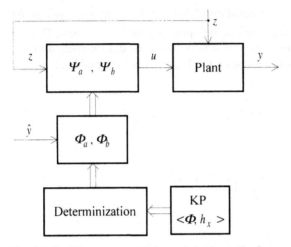

Fig. 2.2. Decision system with determinization – the first case

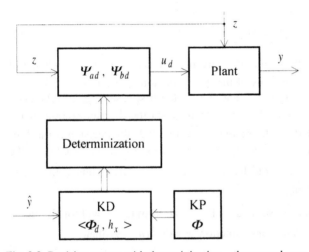

Fig. 2.3. Decision system with determinization – the second case

Example 2.3: Let $u, y, x, z \in R^1$ and

$$y = xu + z .$$

Then

$$M_y(\overline{y}) = u M_x(\overline{x}) + z$$

and from the equation $M_y(\overline{y}) = \hat{y}$ we obtain

$$u_b = \Psi_b(z) = \frac{\hat{y} - z}{M_x(\bar{x})}.$$

The uncertain decision algorithm is

$$u = \Phi_d(z, x) = \frac{\hat{y} - z}{x}$$

and after the determinization

$$u_{bd} = \Psi_{bd}(z) = \frac{\hat{y} - z}{M_x(\bar{x}^{-1})} \neq \Psi_b(z).$$

This very simple example shows that the deterministic decision algorithm $\Psi_b(z)$ obtained via the determinization of the uncertain plant may differ from the deterministic decision algorithm $\Psi_{bd}(z)$ obtained as a result of the determinization of the uncertain decision algorithm.

2.4 Nonparametric Uncertainty

The certainty distribution $h_y(y; u, z)$ may be given directly by an expert as a nonparametric description of the uncertain plant. If u and z are considered as values of uncertain variables $\bar{u}$ and $\bar{z}$, respectively, then

$$h_y(y; u, z) = h_y(y \mid u, z),$$

i.e. $h_y(y \mid u, z)$ is a conditional certainty distribution. If the certainty distribution $h_z(z)$ for $\bar{z}$ is also given by an expert then it is possible to find the uncertain decision algorithm in the form of a conditional certainty distribution $h_u(u \mid z)$, for the given desirable certainty distribution $h_y(y)$ required by a user.

Decision problem: For the given $h_y(y \mid u, z)$, $h_z(z)$ and $h_y(y)$ one should determine $h_u(u \mid z)$.

According to the relationships (1.74), (1.75) and (1.76)

$$h_y(y) = \max_{u \in U, z \in Z} h_y(y, u, z)$$

where $h_y(y, u, z)$ is the joint certainty distribution for $(\bar{y}, \bar{u}, \bar{z})$, i.e.

$$h_y(y, u, z) = \max_{u \in U, z \in Z} \min\{h_{uz}(u, z), h_y(y \mid u, z)\}$$

and the joint certainty distribution

$$h_{uz}(u, z) = \min\{h_z(z), h_u(u \mid z)\}.$$ (2.16)

Finally

$$h_y(y) = \max_{u \in U,\, z \in Z} \min\{h_z(z), h_u(u \mid z), h_y(y \mid z, u)\}.$$ (2.17)

Any distribution $h_u(u \mid z)$ satisfying the equation (2.17) is a solution of our decision problem. It is easy to note that the solution of the equation (2.17) with respect to $h_u(u, z)$ is not unique, i.e. the equation (2.17) may be satisfied by different conditional certainty distributions $h_u(u \mid z)$. Having $h_u(u \mid z)$ one can obtain the deterministic decision algorithm after the determinization of the uncertain decision algorithm described by $h_u(u \mid z)$, according to (2.13) or (2.15) with $h_u(u \mid z)$ instead of $h_u(u; z)$. The decision algorithms $\Psi_{ad}(z)$ or $\Psi_{bd}(z)$ are then obtained as a result of the determinization of the knowledge of the decision making $KD = < h_u(u \mid z) >$, which is determined from the knowledge of the plant

$$KP = < h_y(y \mid u, z), h_z(z) >$$

for the given $h_y(y)$ (Fig. 2.4). It is worth noting that the deterministic decision algorithm obtained in this way has no clear practical interpretation. It is introduced here mainly for the comparison with a fuzzy approach presented in Chap. 6.

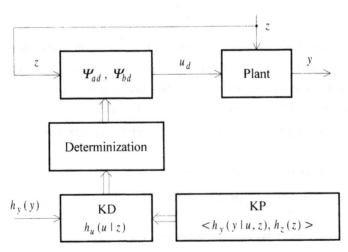

Fig. 2.4. Open-loop decision system under consideration

The determination of $h_u(u \mid z)$ may be decomposed into two steps. In the first step we determine the sets of the joint certainty distributions $h_{uz}(u, z)$ satisfying the equation

$$h_y(y) = \max_{u \in U, z \in Z} \min\{h_{uz}(u, z), h_y(y \mid u, z)\} \qquad (2.18)$$

and in the second step we determine $h_u(u \mid z)$ from equation (2.16). It is easy to see that if the functions $h_y(y)$ and $h_y(y \mid u, z)$ have one local maximum equal to 1 then the point (u, z) maximizing the right hand side of equation (2.18) satisfies the equation

$$h_{uz}(u, z) = h_y(y \mid u, z) .$$

Hence, for this point we have

$$h_y(y) = h_y(y \mid u, z) . \qquad (2.19)$$

Theorem 2.2: The set of functions $h_u(u \mid z)$ satisfying the equation (2.16) is determined as follows:

$$h_u(u \mid z) \begin{cases} = h_{uz}(u, z) & \text{for} \quad (u, z) \notin D(u, z) \\ \geq h_{uz}(u, z) & \text{for} \quad (u, z) \in D(u, z) \end{cases}$$

where

$$D(u, z) = \{(u, z) \in U \times Z : \quad h_z(z) = h_{uz}(u, z)\} .$$

Proof: From (2.16) it follows that

$$\bigwedge_{u \in U} \bigwedge_{z \in Z} [h_z(z) \geq h_{uz}(u, z)] .$$

If $h_z(z) > h_{uz}(u, z)$ then, according to (2.16), $h_{uz}(u, z) = h_u(u \mid z)$. If $h_z(z) = h_{uz}(u, z)$, i.e. $(u, z) \in D(u, z)$ then $h_u(u \mid z) \geq h_{uz}(u, z)$. □

In particular, as one of the solutions of the equation (2.16) we may accept

$$h_u(u \mid z) = h_{uz}(u, z) . \qquad (2.20)$$

Consequently, we may apply the following procedure for the determination of the uncertain decision algorithm:

1. To solve the equation (2.19) with respect to y and to obtain $y^*(u, z)$.

2. To put $y^*(u, z)$ into $h_y(y)$ in the place of y and to obtain

$$h_{uz}(u, z) = h_y[y^*(u, z)].$$

3. To assume $h_u(u \mid z) = h_{uz}(u, z)$.

Let us note that under the assumption (2.20) the knowledge of $h_z(z)$ is not necessary for the determination of the uncertain decision algorithm.

Example 2.4: Consider a plant with $u, y, z \in R^1$, described by the conditional certainty distribution given by an expert:

$$h_y(y \mid u, z) = -(y - d)^2 + 1 - u - (b - z)$$

for

$$0 \le u \le \frac{1}{2}, \quad b - \frac{1}{2} \le z \le b,$$

$$-\sqrt{1 - x - (b - z)} + d \le y \le \sqrt{1 - x - (b - z)} + d,$$

and $h_y(y \mid u, z) = 0$ otherwise.

For the certainty distribution required by a user (Fig. 2.5):

$$h_y(y) = \begin{cases} -(y - c)^2 + 1 & \text{for} \quad c - 1 \le y \le c + 1 \\ 0 & \text{otherwise}, \end{cases}$$

one should determine the uncertain decision algorithm in the form

$$h_u(u \mid z) = h_{uz}(u, z).$$

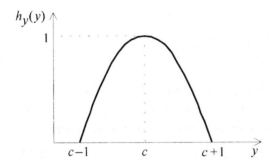

Fig. 2.5. Example of certainty distribution

Let us assume that

$$c + 1 \le d \le c + 2. \tag{2.21}$$

Then the equation (2.19) has a unique solution which is reduced to the solution of the equation

$$-(y-c)^2 + 1 = -(y-d)^2 + 1 - u - (b-z)$$

and

$$y^* = \frac{d^2 - c^2 + u + b - z}{2(d-c)} = \frac{1}{2}(d + c + \frac{u+b-z}{d-c}). \qquad (2.22)$$

Using (2.22) and (2.21) we obtain

$$h_{uz}(u \mid z) = h_{uz}(u, z) = h_y(y^*)$$

$$= \begin{cases} -[\dfrac{(d-c)^2 + u + b - z}{2(d-c)}]^2 + 1 & \text{for} \quad u - z \le 1 - [d - (c+1)]^2 - b, \\ & \quad 0 \le u \le \dfrac{1}{2}, \quad b - \dfrac{1}{2} \le z \le b \\ 0 & \text{otherwise} \quad . \end{cases}$$

3 Relational Systems

3.1 Relational Knowledge Representation

The considerations presented in the previous chapter may be extended for static relational systems, i.e. the systems described by relations which are not reduced to functions. Let us consider a static plant with input vector $u \in U$ and output vector $y \in Y$, where U and Y are real number vector spaces. The plant is described by a relation

$$u \, \rho \, y \overset{\Delta}{=} R(u, y) \subset U \times Y \qquad (3.1)$$

which may be called a *relational knowledge representation* of the plant. It is an extension of the traditional functional model $y = \Phi(u)$ considered in the previous chapter. The description (3.1) given by an expert may have two practical interpretations:

1. The plant is deterministic, i.e. at every moment n

$$y_n = \Phi(u_n),$$

but the expert has no full knowledge of the plant and for the given u he can determine only the set of possible outputs:

$$D_y(u) \subset Y : \{y \in Y : (u, y) \in R(u, y)\}.$$

For example, in one-dimensional case $y = cu$, the expert knows that $c_1 \leq c \leq c_2$; $c_1, c_2 > 0$. Then as the description of the plant he gives a relation presented in the following form

$$\left. \begin{array}{ll} c_1 u \leq y \leq c_2 u & \text{for} \quad u \geq 0 \\ c_2 u \leq y \leq c_1 u & \text{for} \quad u \leq 0 \end{array} \right\} . \qquad (3.2)$$

The situation is illustrated in Fig. 3.1, in which the set of points (u_n, y_n) is denoted.

2. The plant is not deterministic, which means that at different n we may observe

different values y_n for the same values u_n. Then $R(u, y)$ is a set of all possible points (u_n, y_n), denoted for the example (3.2) in Fig. 3.2.

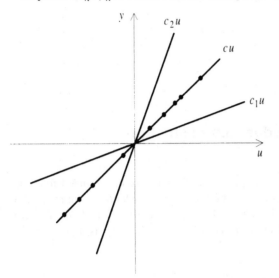

Fig. 3.1. Illustration of a relation – the first case

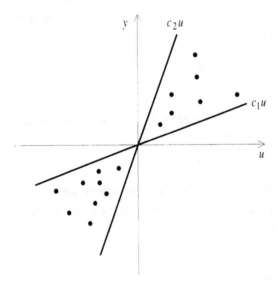

Fig. 3.2. Illustration of a relation – the second case

In the first case the relation (which is not a function) is a result of the expert's uncertainty and in the second case – a result of the uncertainty in the plant. For simplicity, in both cases we shall say about an *uncertain plant*, and the plant described by a relational knowledge representation will be shortly called a *relational plant*.

In more complicated cases the relational knowledge representation given by an expert may have a form of a set of relations:

$$R_i(u, w, y) \subset U \times W \times Y, \quad i = 1, 2, ..., k \tag{3.3}$$

where $w \in W$ is a vector of additional auxiliary variables used in the description of the knowledge. The set of relations (3.3) may be called a *based knowledge representation*. It may be reduced to a *resulting knowledge representation* $R(u, y)$:

$$R(u, y) = \{(u, y) \in U \times Y : \bigvee_{w \in W} (u, w, y) \subset \overline{R}(u, w, y)\}$$

where

$$\overline{R}(u, w, y) = \bigcap_{i=1}^{k} R_i(u, w, y).$$

The relations $R_i(u, w, y)$ may have a form of a set inequalities and/or equalities concerning the components of the vectors u, w, y.

3.2 Analysis and Decision Making for Relational Plants

The formulations of the analysis and decision making problems for a relational plant analogous to those for a functional plant described by a function $y = \Phi(u)$ are adequate to the knowledge of the plant [14].

Analysis problem may be formulated as follows: For the given $R(u, y)$ and $D_u \subset U$ find the smallest set $D_y \subset Y$ such that the implication

$$u \in D_u \rightarrow y \in D_y \tag{3.4}$$

is satisfied.

The information that $u \in D_u$ may be considered as a result of observation. For the given D_u one should determine the best estimation of y in the form of the set of possible outputs D_y. It is easy to note that

$$D_y = \{ y \in Y : \bigvee_{u \in D_u} (u, y) \in R(u, y)\}. \tag{3.5}$$

This is then a set of all such values of y for which there exists $u \in D_u$ such that (u, y) belongs to R. In particular, if the value u is known, i.e. $D_u = \{u\}$

(a singleton) then

$$D_y(u) = \{ y \in Y : (u, y) \in R(u, y) \} \tag{3.6}$$

where $D_y(u)$ is a set of all possible y for the given value u. The analysis problem is illustrated in Fig. 3.3 where the shaded area illustrates the relation $R(u, y)$ and the interval D_y denotes the solution for the given interval D_u.

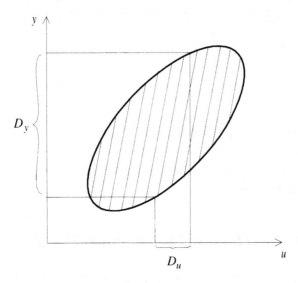

Fig. 3.3. Illustration of analysis problem

Example 3.1: Let us consider the plant with two inputs $u^{(1)}$ and $u^{(2)}$, described by inequality

$$c_1 u^{(1)} + d_1 u^{(2)} \le y \le c_2 u^{(1)} + d_2 u^{(2)} ,$$

and the set D_u is determined by inequalities

$$au^{(1)} + bu^{(2)} \le \alpha \tag{3.7}$$

$$u^{(1)} \ge u^{(1)}_{min}, \qquad\qquad u^{(2)} \ge u^{(2)}_{min} . \tag{3.8}$$

For example, y may denote the amount of a product in a production process, $u^{(1)}$ and $u^{(2)}$ – amounts of two kinds of a raw material, and the value $au^{(1)} + bu^{(2)}$ – a cost of the raw material. Assume that c_1, c_2, d_1, d_2, a, b, α are positive numbers and $c_1 < c_2$, $d_1 < d_2$. It is easy to see that the set (3.5) is described by inequality

$$c_1 u_{min}^{(1)} + d_1 u_{min}^{(2)} \leq y \leq y_{max} \tag{3.9}$$

where

$$y_{max} = \max_{u^{(1)}, u^{(2)}} (c_2 u^{(1)} + d_2 u^{(2)}) \tag{3.10}$$

subject to constraints (3.7) and (3.8).

The maximization in (3.10) leads to the following results:

If

$$\frac{c_2}{d_2} \leq \frac{a}{b}$$

then

$$y_{max} = c_2 u_{min}^{(1)} + \frac{d_2}{b} (\alpha - a u_{min}^{(1)}) . \tag{3.11}$$

If

$$\frac{c_2}{d_2} \geq \frac{a}{b}$$

then

$$y_{max} = \frac{c_2}{a} (\alpha - b u_{min}^{(2)}) + d_2 u_{min}^{(2)} . \tag{3.12}$$

For the numerical data $c_1 = 1$, $c_2 = 2$, $d_1 = 2$, $d_2 = 4$, $a = 1$, $b = 4$, $\alpha = 3$, $u_{min}^{(1)} = 1$, $u_{min}^{(2)} = 0.5$

$$\frac{c_2}{d_2} = \frac{1}{2}, \qquad \frac{a}{b} = \frac{1}{4} < \frac{c_2}{d_2} .$$

From (3.12) we obtain $y_{max} = 4$ and according to (3.9) $y_{min} = c_1 u_{min}^{(1)} + d_1 u_{min}^{(2)} = 2$. The set D_y is then determined by inequality $2 \leq y \leq 4$.   □

Decision problem: For the given $R(u, y)$ and $D_y \subset Y$ find the largest set $D_u \subset U$ such that the implication (3.4) is satisfied.

The set D_y is given by a user, the property $y \in D_y$ denotes the user's requirement and D_u denotes the set of all possible decisions for which the requirement concerning the output y is satisfied. It is easy to note that

$$D_u = \{u \in U : \quad D_y(u) \subseteq D_y\} \tag{3.13}$$

where $D_y(u)$ is the set of all possible y for the fixed value u, determined by (3.6). The solution may not exist, i.e. $D_u = \varnothing$ (empty set). In the example illustrated in Fig. 3.2, if $D_y = [y_{min}, y_{max}]$ and $c_1, c_2 > 0$ then

$$D_u = [\frac{y_{min}}{c_1}, \frac{y_{max}}{c_2}]$$

and the solution exists under the condition

$$\frac{y_{min}}{c_1} \leq \frac{y_{max}}{c_2} .$$

The analysis and decision problems for the relational plant are the extensions of the respective problems for the functional plant, presented in Sect. 2.1. The properties "$u \in D_u$" and "$y \in D_y$" may be called the *input and output properties*, respectively. For the functional plant we considered the input and output properties in the form: "$u = u^*$" and "$y = y^*$" where u^*, y^* denote fixed variables. For the relational plant the analysis problem consists in finding the best output property (the smallest set D_y) for the given input property, and the decision problem consists in finding the best input property (the largest set D_u) for the given required output property. The procedure for determining the effective solution D_u and D_y based on the general formulas (3.5) and (3.13) depends on the form of $R(u, y)$ and may be very complicated. If $R(u, y)$ and the given property (i.e. the given set D_u or D_y) are described by a set of equalities and/or inequalities concerning the components of the vector u and y then the procedure is reduced to "solving" this set of equalities and/or inequalities.

Example 3.2: Consider a plant described by a relation

$$G_1(u) \leq y \leq G_2(u) \tag{3.14}$$

where G_1 and G_2 are the functions

$$G_1 : U \to R^+, \qquad\qquad G_2 : U \to R^+; \qquad\qquad R^+ = [0, \infty),$$

and

$$\bigwedge_{u \in U} [G_1(u) \leq G_2(u)].$$

For example, y is the amount of a product as in Example 3.1 and the components of the vector u are features of the raw materials. For a user's requirement

$$y_{min} \le y \le y_{max},$$

i.e. $D_y = [y_{min}, y_{max}]$, we obtain

$$D_u = \{u \in U : [G_1(u) \ge y_{min}] \wedge [G_2(u) \le y_{max}]\}.$$

In particular, if the relation (3.14) has a form

$$c_1 u^T u \le y \le c_2 u^T u, \qquad c_1 > 0, c_2 > c_1 \tag{3.15}$$

where $u \in R^k$ and

$$u^T u = (u^{(1)})^2 + (u^{(2)})^2 + ... + (u^{(k)})^2$$

then D_u is described by inequality

$$\frac{y_{min}}{c_1} \le u^T u \le \frac{y_{max}}{c_2}$$

and the decision u satisfying the requirement (3.15) exists iff

$$\frac{y_{max}}{c_2} \ge \frac{y_{min}}{c_1}. \qquad \qquad \square$$

The considerations may by extended for a plant with external disturbances, described by a relation $R(u, y, z) \subset U \times Y \times Z$ where $z \in Z$ is a vector of the disturbances which may be observed. The property $z \in D_z$ for the given $D_z \subset Z$ may be considered as a result of the observations. Our plant has two inputs (u, z) and analysis problem is formulated in the same way as for the relation $R(u, y)$, with $(u, z) \in D_u \times D_z$ in the place of $u \in D_u$. The result analogous to (3.5) is

$$D_y = \{y \in Y : \bigvee_{u \in D_u} \bigvee_{z \in D_z} (u, y, z) \in R(u, y, z)\}.$$

Decision problem: For the given $R(u, y, z)$, D_y (the requirement) and D_z (the result of observations), find the largest set D_u such that the implication

$$(u \in D_u) \wedge (z \in D_z) \rightarrow y \in D_y$$

is satisfied. The general form of the solution is as follows

$$D_u = \{u \in U : \bigwedge_{z \in D_z} [D_y(u,z) \subseteq D_y]\} \tag{3.16}$$

where

$$D_y(u,z) = \{y \in Y : (u,y,z) \in R(u,y,z)\} . \tag{3.17}$$

It is then the set of all such decisions u that for every $z \in D_z$ the set of possible outputs y belongs to D_y. For the fixed z (the result of measurement) the set D_u is determined by (3.16) with the relation

$$R(u,y,z) \stackrel{\Delta}{=} R(u,y;z) \subset U \times Y .$$

In this notation z is the parameter in the relation $R(u,y;z)$. Then

$$D_u(z) = \{u \in U : D_y(u,z) \subseteq D_y\} \stackrel{\Delta}{=} \overline{R}(z,u) \tag{3.18}$$

where $D_y(u,z)$ is defined by (3.17). The formula (3.18) defines a relation between z and u denoted by $\overline{R}(z,u)$. The relation $\overline{R}(z,u)$ may be called a knowledge representation for the decision making (the description of the knowledge of the decision making) or a *relational decision algorithm*. The block scheme of the open-loop decision system (Fig. 3.4) is analogous to that in Fig. 2.1 for a functional plant. The *knowledge of the decision making*

$$< \overline{R}(z,u) > \stackrel{\Delta}{=} KD$$

has been obtained for the given *knowledge of the plant*

$$< R(u,y,z) > \stackrel{\Delta}{=} KP$$

and the given requirement $y \in D_y$.

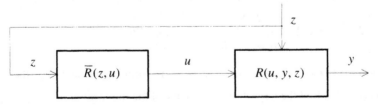

Fig. 3.4. Open-loop decision system

3.3 Determinization

The deterministic decision algorithm based on the knowledge KD may be obtained as a result of determinization of the relational decision algorithm $\overline{R}(z,u)$ by using the mean value

$$\tilde{u}(z) = \int\limits_{D_u(z)} u\,du \cdot [\int\limits_{D_u(z)} du\,]^{-1} \triangleq \tilde{\Psi}(z).$$

For the given desirable value $\hat{y}$ we can consider two cases analogous to the concepts described in Sect. 2.3 and illustrated in Fig. 2.2 and Fig. 2.3. In the first case the deterministic decision algorithm $\Psi(z)$ is obtained via the determinization of the knowledge of the plant KP and in the second case the deterministic decision algorithm $\Psi_d(z)$ is based on the determinization of the knowledge of the decision making KD obtained from KP for the given $\hat{y}$. In the first case we determine the mean value

$$\tilde{y}(z) = \int\limits_{D_y(u,z)} y\,dy \cdot [\int\limits_{D_y(u,z)} dy\,]^{-1} \triangleq \Phi(u,z) \tag{3.19}$$

where $D_y(u,z)$ is described by formula (3.17). Then, solving the equation

$$\Phi(u,z) = \hat{y} \tag{3.20}$$

with respect to u, we obtain the deterministic decision algorithm $u = \Psi(z)$, under the assumption that the equation (3.20) has a unique solution.

In the second case we use

$$R(u,\hat{y},z) \triangleq R_d(z,u), \tag{3.21}$$

i.e. the set of all pairs (u,z) for which it is possible that $y = \hat{y}$. The relation $R_d(z,u) \subset Z \times U$ may be considered as the knowledge of the decision making KD, i.e. the relational decision algorithm obtained for the given KP and the value $\hat{y}$. The determinization of the relational decision algorithm R_d gives the deterministic decision algorithm:

$$u_d(z) = \int\limits_{D_{ud}(z)} u\,du \cdot [\int\limits_{D_{ud}(z)} du\,]^{-1} \triangleq \Psi_d(z) \tag{3.22}$$

where

$$D_{ud}(z) = \{u \in U : (u, z) \in R_d(z, u)\}.$$

The equations (3.19), (3.20), (3.22) are analogous to the equations (2.6), (2.9), (2.15) presented in Sect. 2.3. Two cases of the determination of the deterministic decision algorithm are illustrated in Figs. 3.5 and 3.6, analogous to Figs. 2.2 and 2.3. The results of these two approaches may be different, i.e. in general $\Psi(z) \neq \Psi_d(z)$ (see Example 3.3).

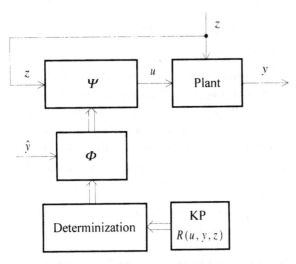

Fig. 3.5. Decision system with determinization – the first case

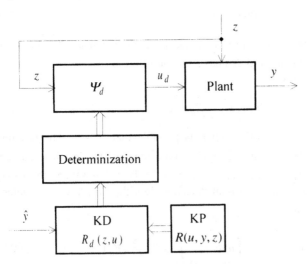

Fig. 3.6. Decision system with determinization – the second case

Example 3.3:   Consider the plant with $u, z, y \in R^1$ (one-dimensional variables), described by the inequality

$$cu + z \le y \le 2cu + z , \qquad c > 0 . \tag{3.23}$$

For $D_y = [y_{min}, y_{max}]$ and the given z the set (3.18) is determined by the inequality

$$\frac{y_{min} - z}{c} \le u \le \frac{y_{max} - z}{2c} .$$

The determinization of the knowledge KP according to (3.19) gives

$$\tilde{y} = \frac{3}{2} cu + z = \Phi(u, z) .$$

From the equation $\Phi(u, z) = \hat{y}$ we obtain the decision algorithm

$$u = \Psi(z) = \frac{2(\hat{y} - c)}{3z} .$$

Substituting $\hat{y}$ into (3.23) we obtain the relational decision algorithm $R_d(z, u)$ in the form

$$\frac{\hat{y} - z}{2c} \le u \le \frac{\hat{y} - z}{c}$$

and after the determinization

$$u_d = \Psi_d(z) = \frac{3(\hat{y} - z)}{4c} \ne \Psi(z) .$$

3.4 Analysis for Relational Plants with Uncertain Parameters [26, 35]

Let us consider the plant described by a relation $R(u, y ; x) \subseteq U \times Y$ where $x \in X$ is an unknown vector parameter which is assumed to be a value of an uncertain variable $\bar{x}$ with the certainty distribution $h_x(x)$ given by an expert. Now the sets of all possible values y in (3.5) and (3.6) depend on x. For the given set of inputs D_u we have

$$D_y(x) = \{ y \in Y : \bigvee_{u \in D_u} (u, y) \in R(u, y ; x) \}$$

and for the given value u

$$D_y(u\,;x) = \{y \in Y :\ (u, y) \in R(u, y\,;x)\}\,.$$

The analysis may consist in evaluating the input with respect to a set $D_y \subset Y$ given by a user. We can consider two formulations with the different practical interpretation: the determination of $v[D_y \,\tilde{\subseteq}\, D_y(\bar{x})]$ (version I) or the determination of $v[D_y(\bar{x}) \subseteq D_y]$ (version II). The analogous formulations may be considered for the given u, with $D_y(u\,;\bar{x})$ in the place of $D_y(\bar{x})$.

Analysis problem – version I: For the given $R(u, y\,;x)$, $h_x(x)$, u and $D_y \subset Y$ one should determine

$$v[D_y \,\tilde{\subseteq}\, D_y(u\,;\bar{x})] \triangleq g(D_y,u)\,. \tag{3.24}$$

The value (3.24) denotes the certainty index of the soft property: "the set of all possible outputs approximately contains the set D_y given by a user" or "the approximate value of $\bar{x}$ is such that $D_y \subseteq D_y(u\,;x)$" or "the approximate set of the possible outputs contains all the values from the set D_y". Let us note that

$$v[D_y \,\tilde{\subseteq}\, D_y(u\,;\bar{x})] = v[\bar{x} \,\tilde{\in}\, D_x(D_y,u)] \tag{3.25}$$

where

$$D_x(D_y,u) = \{x \in X :\ D_y \subseteq D_y(u\,;x)\}\,. \tag{3.26}$$

Then

$$g(D_y,u) = \max_{x \in D_x(D_y,u)} h_x(x)\,. \tag{3.27}$$

In particular, for $D_y = \{y\}$ (a singleton), the certainty index that the given value y may occur at the output of the plant is

$$g(y,u) = \max_{x \in D_x(y,u)} h_x(x)\,. \tag{3.28}$$

where

$$D_x(y,u) = \{x \in X :\ y \in D_y(u\,;x)\}\,. \tag{3.29}$$

When $\bar{x}$ is considered as C-uncertain variable, it is necessary to determine

$$v[\bar{x} \,\tilde{\in}\, \overline{D}_x(D_y,u)] = \max_{x \in \overline{D}_x(D_y,u)} h_x(x)$$

where $\overline{D}_x(D_y, u) = X - D_x(D_y, u)$. Then, according to (1.59)

$$v_c[D_y \tilde{\subseteq} D_y(u; \bar{x})] = \frac{1}{2}\{v[\bar{x} \tilde{\in} D_x(D_y, u)] + 1 - v[\bar{x} \tilde{\in} \overline{D}_x(D_y, u)]\}.$$

The considerations may be extended for a plant described by a relation $R(u, y, z; x)$ where $z \in Z$ is the vector of disturbances which may be measured. For the given z

$$D_y(u, z; x) = \{y \in Y : \quad (u, y, z) \in R(u, y, z; x)\}$$

and

$$v[D_y \tilde{\subseteq} D_y(u, z; \bar{x})] = \max_{x \in D_x(D_y, u, z)} h_x(x)$$

where

$$D_x(D_y, u, z) = \{x \in X : \quad D_y \subseteq D_y(u, z; x)\}.$$

Consequently, the certainty index that the approximate set of the possible outputs contains all the values from the set D_y depends on z.

For the given set D_u, the formulas analogous to (3.24) – (3.29) have the following form:

$$v[D_y \tilde{\subseteq} D_y(\bar{x})] \stackrel{\Delta}{=} g(D_y, D_u),$$

$$v[D_y \tilde{\subseteq} D_y(\bar{x})] = v[\bar{x} \tilde{\in} D_x(D_y, D_u)],$$

$$D_x(D_y, D_u) = \{x \in X : \quad D_y \subseteq D_y(x)\},$$

$$g(D_y, D_u) = \max_{x \in D_x(D_y, D_u)} h_x(x),$$

$$g(y, D_u) = \max_{x \in D_x(y, D_u)} h_x(x),$$

$$D_x(y, D_u) = \{x \in X : \quad y \in D_y(x)\}.$$

Analysis problem – version II: For the given $R(u, y; x)$, $h_x(x)$, u and $D_y \subset Y$ one should determine

$$v[D_y(u; \bar{x}) \tilde{\subseteq} D_y] \stackrel{\Delta}{=} g(D_y, u). \tag{3.30}$$

The value (3.30) denotes the certainty index of the soft property: "the set D_y

given by a user contains the approximate set of all possible outputs". The formulas corresponding to (3.25), (3.26) and (3.27) are as follows:

$$v[D_y(u;\bar{x}) \tilde{\subseteq} D_y] = v[\bar{x} \in D_x(D_y,u)]$$

where

$$D_x(D_y,u) = \{x \in X : \ D_y(u;x) \subseteq D_y\}, \tag{3.31}$$

$$g(D_y,u) = \max_{x \in D_x(D_y,u)} h_x(x). \tag{3.32}$$

For the given set D_u one should determine

$$v[D_y(\bar{x}) \tilde{\subseteq} D_y] = v[\bar{x} \tilde{\in} D_x(D_y,D_u)] = \max_{x \in D_x(D_y,D_u)} h_x(x) \tag{3.33}$$

where

$$D_x(D_y,D_u) = \{x \in X : \ D_y(x) \subseteq D_y\}. \tag{3.34}$$

In the case where $\bar{x}$ is considered as C-uncertain variable it is necessary to find v (3.33) and

$$v[\bar{x} \tilde{\in} \overline{D}_x(D_y,D_u)] = \max_{x \in \overline{D}_x(D_y,D_u)} h_x(x) \tag{3.35}$$

where $\overline{D}_x(D_y,D_u) = X - D_x(D_y,D_u)$. Then, according to (1.59)

$$v_c[D_y(\bar{x}) \tilde{\subseteq} D_y] = \frac{1}{2}\{v[\bar{x} \tilde{\in} D_x(D_y,D_u)] + 1 - v[\bar{x} \tilde{\in} \overline{D}_x(D_y,D_u)]\}. \tag{3.36}$$

The considerations for the plant described by $R(u,y,z;x)$ are analogous to those in version I.

Example 3.4: Let u, y, $x \in R^1$, the relation R is given by inequality $xu \le y \le 2xu$, $D_u = [u_1, u_2]$, $u_1 > 0$, $D_y = [y_1, y_2]$, $y_1 > 0$. For these data $D_y(x) = [xu_1, 2xu_2]$ and (3.34) becomes $D_x(D_y,D_u) = [\frac{y_1}{u_1}, \frac{y_2}{2u_2}]$. Assume that x is a value of an uncertain variable $\bar{x}$ with triangular certainty distribution: $h_x = 2x$ for $0 \le x \le \frac{1}{2}$, $h_x = -2x + 2$ for $\frac{1}{2} \le x \le 1$, $h_x = 0$ otherwise (Fig. 3.7). Using (3.33) we obtain for $u_1 y_2 \ge 2u_2 y_1$

$$v[D_y(\bar{x}) \widetilde{\subseteq} D_y] = v[\bar{x} \widetilde{\in} D_x(D_y, D_u)]$$

$$= \begin{cases} \dfrac{y_2}{u_2} & \text{when} & y_2 \leq u_2 \\ 1 & \text{when} & 2y_1 \leq u_1 \text{ and } y_2 \geq u_2 \\ 2\left(1 - \dfrac{y_1}{u_1}\right) & \text{when} & 2y_1 \geq u_1 \text{ and } y_1 \leq u_1 \\ 0 & \text{when} & y_1 \geq u_1 . \end{cases}$$

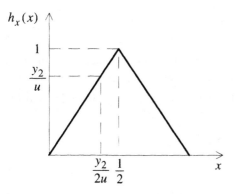

Fig. 3.7. Example of certainty distribution

For $u_1 y_2 < 2u_2 y_1$ $D_x(D_y, D_u) = \varnothing$ and $v[D_y(\bar{x}) \widetilde{\subseteq} D_y] = 0$. For example, for $u_1 = 5$, $u_2 = 6$, $y_1 = 4$, $y_2 = 12$ we have $2y_1 \geq u_1$, $y_1 \leq u_1$ and $v(D_y(\bar{x}) \widetilde{\subseteq} [4, 12]) = 0.4$. When $y_1 = 2$ we have $2y_1 \leq u_1$, $y_2 \geq u_2$ and $v = 1$. To apply the description for C-uncertain variable one should determine $v[\bar{x} \in \overline{D}_x(D_y, D_u)]$ according to (3.35):

$$v[\bar{x} \in \overline{D}_x(D_y, D_u)] = \begin{cases} 1 & \text{when} & y_2 \leq u_2 \\ \max\left(\dfrac{2y_1}{u_1}, \dfrac{-y_2}{u_2} + 2\right) & \text{when} & 2y_1 \leq u_1 \text{ and } y_2 \geq u_2 \\ 1 & \text{when} & 2y_1 \geq u_1 . \end{cases}$$

Then, using (3.36) we obtain $v_c[D_y(\bar{x}) \widetilde{\subseteq} D_y]$. For the numerical data in the first case ($y_1 = 4$) $v[\bar{x} \in \overline{D}_x(D_y, D_u)] = 1$, $v_c = 0.2$. In the second case ($y_1 = 2$) $v[\bar{x} \in \overline{D}_x(D_y, D_u)] = 0.8$, $v_c = 0.6$.

3.5 Decision Making for Relational Plants with Uncertain Parameters [35, 36]

We can formulate the different versions of the decision problem with different practical interpretations – corresponding to the formulations of the analysis problem presented in Sect. 3.4.

Decision problem – version I: For the given $R(u, y; x)$, $h_x(x)$ and $D_y \subset Y$ find

$$\hat{u} = \arg \max_{u \in U} \; v[D_y \tilde{\subseteq} D_y(u; \bar{x})]. \tag{3.37}$$

In this formulation $\hat{u}$ is a decision maximizing the certainty index that the approximate set of the possible outputs contains the set D_y given by a user. To obtain the optimal decision one should determine the function g in (3.27) and to maximize it with respect to u, i.e.

$$\hat{u} = \arg \max_{u \in U} \; \max_{x \in D_x(D_y, u)} h_x(x) \tag{3.38}$$

where $D_x(D_y, u)$ is defined by (3.26).

Decision problem – version II: For the given $R(u, y; x)$, $h_x(x)$ and $D_y \subset Y$ find

$$\hat{u} = \arg \max_{u \in U} \; v[D_y(u; \bar{x}) \tilde{\subseteq} D_y]. \tag{3.39}$$

Now $\hat{u}$ is a decision maximizing the certainty index that the approximate set of all possible outputs (i.e. the set of all possible outputs for the approximate value of $\bar{c}$) belongs to the set D_y given by a user. To obtain the optimal decision one should determine the function g in (3.32) and to maximize it with respect to u, i.e. $\hat{u}$ is determined by (3.38) where $D_x(D_y, u)$ is defined by (3.31). It is worth noting that in both versions the solution may not be unique, i.e. we may obtain the set D_u of the decisions (3.37). Denote by x^* the value maximizing $h_x(x)$, i.e. such that $h_x(x^*) = 1$. Then

$$D_u = \{u \in U : \; x^* \in D_x(D_y, u)\}$$

and for every $u \in D_u$ maximal value of the certainty index in (3.37) and (3.39) is equal to 1. To determine the set of optimal decisions D_u it is not necessary to know the form of the function $f_x(x)$. It is sufficient to know the value x^*.

In the case where $\bar{x}$ is considered as C-uncertain variable one should determine

$$v_c[\bar{x} \tilde{\in} D_x(D_y,u)] = \frac{1}{2}\{v[\bar{x} \tilde{\in} D_x(D_y,u)] + 1 - v[\bar{x} \tilde{\in} \bar{D}_x(D_y,u)]\} \qquad (3.40)$$

where

$$v[\bar{x} \tilde{\in} \bar{D}_x(D_y,u)] = \max_{x \in \bar{D}_x(D_y,u)} h_x(x). \qquad (3.41)$$

Then the optimal decision $\hat{u}_c$ is obtained by maximization of v_c :

$$\hat{u}_c = \max_{u \in U} v_c[\bar{x} \tilde{\in} D_x(D_y,u)]$$

where $D_x(D_y,u)$ is defined by (3.26) in version I or by (3.31) in version II.

In the similar way as in Sect. 3.4, the considerations may be extended for the plant with the vector of external disturbances z, described by $R(u, y, z; x)$. Now the set $D_u(z)$ of the optimal decisions depends on z. In the case of the unique solution $\hat{u}$ for every z, we obtain the deterministic decision algorithm $\hat{u} = \Psi(z)$ in an open-loop decision system. It is the decision algorithm based on the knowledge of the plant $KP = < R, h_x >$. For the fixed x and z we may solve the decision problem such as in Sect. 3.2, i.e. determine the largest set $D_u(z; x)$ such that the implication

$$u \in D_u(z; x) \rightarrow y \in D_y$$

is satisfied. According to (3.18)

$$D_u(z; x) = \{u \in U : D_y(u, z; x) \subseteq D_y\} \stackrel{\Delta}{=} \bar{R}(z, u; x)$$

where

$$D_y(u, z; x) = \{y \in Y : (u, y, z) \in R(u, y, z; x)\}.$$

Then we can determine the optimal decision

$$u_d = \arg \max_{u \in U} v[u \tilde{\in} D_u(z; \bar{x})] \stackrel{\Delta}{=} \Psi_d(z) \qquad (3.42)$$

where

$$v[u \tilde{\in} D_u(z; \bar{x})] = v[\bar{x} \tilde{\in} D_{xd}(D_y, u, z)]$$

and

$$D_{xd}(D_y, u, z) = \{x \in X : u \in D_u(z; x)\}.$$

Hence

$$v[u \tilde{\in} D_u(z;\bar{x})] = \max_{x \in D_{xd}(D_y, u, z)} h_x(x). \tag{3.43}$$

In general, we may obtain the set D_{ud} of decisions u_d maximizing the certainty index (3.43). Let us note that the decision algorithm $\Psi_d(z)$ is based on the knowledge of the decision making $KD = <\bar{R}, h_x>$. The relation $\bar{R}$ or the set $D_u(z;x)$ may be called an *uncertain decision algorithm* in the case under consideration. It is easy to see that in this case $u_d = \hat{u}$ for every z, i.e. $\Psi_d(z) = \Psi(z)$ where $\hat{u} = \Psi(z)$ is the optimal decision in version II. This follows from the fact that

$$u \in D_u(z;x) \leftrightarrow D_y(u, z; x) \subseteq D_y,$$

i.e. the properties $u \in D_u(z;x)$ and $D_y(u, z; x) \subseteq D_y$ are equivalent. The optimal decision in version II $\hat{u} = u_d$ is then the decision which with the greatest certainty index belongs to the set of decisions $D_u(z;x)$ for which the requirement $y \in D_y$ is satisfied. The determination of $\hat{u} = u_d$ from (3.42) and (3.43) may be easier than from (3.39) with $D_y(u, z; \bar{x})$ in the place of $D_y(u; \bar{x})$. In the case without z the optimal decision (3.39) may be obtained in the following way:

$$\hat{u} = \arg \max_u v[u \tilde{\in} D_u(\bar{x})]$$

where

$$v[u \tilde{\in} D_u(\bar{x})] = v[\bar{x} \tilde{\in} D_{xd}(D_y, u)] = \max_{x \in D_{xd}(D_y, u)} h_x(x), \tag{3.44}$$

and

$$\left. \begin{array}{l} D_{xd}(D_y, u) = \{x \in X : u \in D_u(x)\}, \\ D_u(x) = \{u \in U : D_y(u; x) \subseteq D_y\}, \\ D_y(u; x) = \{y \in Y : (u, y) \in R(u, y; x)\}. \end{array} \right\} \tag{3.45}$$

Example 3.5 (decision problem – version II): Let $u, y, x \in R^1$ and $R(u, y, x)$ be given by the inequality

$$3x - u \le y \le u^2 + x^2 + 1.$$

For $D_y = [0, 2]$ the set $D_u(x)$ (3.45) is determined by

$$u \le 3x \qquad \text{and} \qquad u^2 + x^2 \le 1. \qquad (3.46)$$

Assume that x is a value of an uncertain variable $\bar{x}$ with triangular certainty distribution: $h_x = 2x$ for $0 \le x \le \frac{1}{2}$, $h_x = -2x + 2$ for $\frac{1}{2} \le x \le 1$, $h_x = 0$ otherwise. From (3.46) we have $D_x(u) = [\frac{u}{3}, \sqrt{1 - u^2}]$ and the set of all possible u: $\Delta_u = [-1, \frac{3}{\sqrt{10}}]$ (the value $\frac{3}{\sqrt{10}}$ is obtained from the equations $u = 3x$, $u^2 + x^2 = 1$). It is easy to see that $\frac{1}{2} \in D_x(u)$ iff $\sqrt{1 - u^2} \ge \frac{1}{2}$. Then, according to (3.44)

$$v[u \tilde{\in} D_u(\bar{x})] \overset{\triangle}{=} v(u) = \begin{cases} 1 & \text{for} \quad -\frac{\sqrt{3}}{2} \le u \le \frac{\sqrt{3}}{2} \\ 2\sqrt{1 - u^2} & \text{otherwise in } \Delta_u. \end{cases} \qquad (3.47)$$

For example $v(0.5) = 1$, $v(0.9) \approx 0.88$. As the decision $\hat{u}$ we can choose any value from $[-\frac{\sqrt{3}}{2}, \frac{\sqrt{3}}{2}]$ and the property $D_y(u; \bar{x}) \tilde{\subseteq} D_y$ is satisfied with certainty index equal to 1. To apply the description for C-uncertain variable it is necessary to determine $v[\bar{x} \tilde{\in} \bar{D}_x(D_y, u)]$. Using (3.41) let us note that for

$$|u| < \frac{\sqrt{3}}{2} \quad v[\bar{x} \tilde{\in} \bar{D}_x(D_y, u)] = \max\{\frac{2u}{3}, \ 2 - 2\sqrt{1 - u^2}\}. \text{ Then}$$

$$v[\bar{x} \tilde{\in} \bar{D}_x(D_y, u)] = \begin{cases} \max\{\frac{2u}{3}, \ 2 - 2\sqrt{1 - u^2}\} & \text{for} \quad -\frac{\sqrt{3}}{2} \le u \le \frac{\sqrt{3}}{2} \\ 1 & \text{otherwise in} \quad \Delta_u. \end{cases}$$
$$(3.48)$$

Substituting (3.47) and (3.48) into (3.40) we obtain $v_c(u)$. For example $v_c(0.5) = \frac{5}{6}$, $v_c(0.9) \approx 0.44$. It is easy to note that in this case $\hat{u}_c = 0$ and $v_c(\hat{u}_c) = 1$. $\qquad\qquad\qquad\qquad\qquad\qquad\qquad\qquad\qquad\qquad\qquad\qquad\quad\square$

Example 3.6 (decision problem – version II): R and $h_x(x)$ are the same as in Example 3.4, $D_y = [y_1, y_2]$, $y_1 > 0$, $y_2 > 2y_1$. Then $D_u(x) = [\frac{y_1}{x}, \frac{y_2}{2x}]$, $D_x(u) = [\frac{y_1}{u}, \frac{y_2}{2u}]$ and $v(u)$ in (3.44) is the same as $v[D_y(\bar{x}) \tilde{\subseteq} D_y]$ in Example 3.4, with $u_1 = u_2 = u$. Thus, $\hat{u}$ is any value from $[2y_1, y_2]$ and $v(\hat{u}) = 1$. In the case of C-uncertain variable $v[\bar{x} \in \overline{D}_x(D_y, u)]$ is the same as in Example 3.4, with $u_1 = u_2 = u$. Using (3.40) we obtain

$$v_c(u) = \begin{cases} \dfrac{y_2}{2u} & \text{when} & u \geq y_1 + 0.5 y_2 \\ 1 - \dfrac{y_1}{u} & \text{when} & y_1 \leq u \leq y_1 + 0.5 y_2 \\ 0 & \text{when} & u \leq y_1 . \end{cases}$$

It is easy to see that $\hat{u}_c = y_1 + 0.5 y_2$ and $v_c(\hat{u}_c) = \dfrac{y_2}{2y_1 + y_2}$. For example, for $y_1 = 2$, $y_2 = 12$ we obtain $\hat{u} \in [4, 12]$ and $v = 1$, $\hat{u}_c = 8$ and $v_c = 0.75$. The function $v_c(u)$ is illustrated in Fig. 3.8.

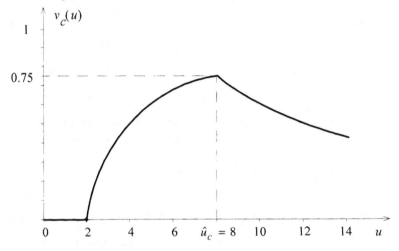

Fig. 3.8. Example of the relationship between v_c and u

3.6 Computational Aspects

The application of C-uncertain variables with the certainty index v_c instead of v means better using the expert's knowledge, but may be connected with much greater computational difficulties. In the discrete case, when the number of possible values x is small, it may be acceptable to determine all possible values of v_c and then to choose the value $\hat{u}_c$ for which v_c is the greatest. Let us explain it for the decision problem in version II. Assume that X and U are finite discrete sets:

$$X = \{x_1, x_2, \dots, x_m\}, \qquad\qquad U = \{u_1, u_2, \dots, u_p\}.$$

Now the relation $R(u, y; x)$ is reduced to the family of sets

$$D_y(u_i; x_j) \subset Y, \qquad\qquad i \in \overline{1, p}, \quad j \in \overline{1, m},$$

i.e. the sets of possible outputs for all the pairs (u_i, x_j).

The algorithm for the determination of $\hat{u}$ is as follows:
1. For u_i $(i = 1, 2, \dots, p)$ prove if

$$D_y(u_i; x_j) \subseteq D_y, \qquad\qquad j = 1, 2, \dots, m \qquad\qquad (3.49)$$

2. Determine

$$v_i = \max_{x \in D_x(D_y, u_i)} h_x(x)$$

where $D_x(D_y, u_i)$ is the set of all x_j satisfying the property (3.49)

3. Choose $\hat{u} = u_i$ for $i = i^*$ where i^* is an index for which v_i is the greatest.

The algorithm for the determination of $\hat{u}_c$ is then the following:
1) For u_i $(i = 1, 2, \dots, p)$ prove if

$$D_y(u_i; x_j) \subseteq D_y, \qquad\qquad j = 1, 2, \dots, m.$$

If yes then $x_j \in D_x(D_y, u_i)$. In this way, for $j = m$ we obtain the set $D_x(D_y, u_i)$ as a set of all x_j satisfying the property (3.49).

2) Determine v_{ci} according to (1.66) and (3.40):

$$v_{ci} = \begin{cases} \dfrac{1}{2} \displaystyle\max_{x \in D_x(D_y, u_i)} h_x(x) & \text{if } x^* \in \overline{D}_x(D_y, u_i) \\[2ex] 1 - \dfrac{1}{2} \displaystyle\max_{x \in \overline{D}_x(D_y, u_i)} h_x(x) & \text{otherwise} \end{cases}$$

where $x^* \in X$ is such that $h_x(x^*) = 1$ and $\overline{D}_x(D_y, u_i) = X - D_x(D_y, u_i)$.

3) Choose $i = i^*$ such that v_{ci} is the maximum value in the set of v_{ci} determined in the former steps. Then $\hat{u}_c = u_i$ for $i = i^*$.

Let us consider the relational plant with one-dimensional output, described by the inequality

$$\Phi_1(u;e) \le y \le \Phi_2(u;d)$$

where $\Phi_1 : U \to \mathbf{R}^1$, $\Phi_2 : U \to \mathbf{R}^1$, e and d are the subvectors of the parameter vector $x = (e,d)$,

$$e \in E = \{e_1, e_2, ..., e_s\}, \qquad d \in D = \{d_1, d_2, ..., d_l\}.$$

Now $m = s \cdot l$ where m is a number of the pairs (e_γ, d_δ); $\gamma \in \overline{1,s}$, $\delta \in \overline{1,l}$. If $D_y = [y_{min}, y_{max}]$ then the set $D_y(u_i; e_\gamma, d_\delta)$ is described by the inequalities

$$\Phi_1(u_i; e_\gamma) \ge y_{min} \qquad \text{and} \qquad \Phi_2(u_i; d_\delta) \le y_{max}.$$

Assume that $\overline{e}$ and $\overline{d}$ are independent uncertain variables. Then, according to (1.76)

$$h_x(x) = h(e,d) = \min\{h_e(e), h_d(d)\}.$$

Let $e^* = e_v$ and $d^* = d_\mu$, i.e. $h_e(e_v) = 1$ and $h_d(d_\mu) = 1$.

The algorithm for the determination of the optimal decision $\hat{u}_c$ in this case is as follows:
1) For u_i prove if

$$\Phi_1(u_i; e_v) \ge y_{min} \qquad \text{and} \qquad \Phi_2(u_i; d_\mu) \le y_{max}.$$

If yes, go to 2). If not got to 4).
2) Prove if

$$\Phi_1(u_i; e_\gamma) \le y_{min} \qquad \text{or} \qquad \Phi_2(u_i; d_\delta) \ge y_{max} \qquad (3.50)$$

for

$$\gamma = 1, 2, ..., v - 1, v + 1, ..., s,$$

$$\delta = 1, 2, ..., \mu - 1, \mu + 1, ..., l.$$

3) Determine

$$v_{ci} = 1 - \frac{1}{2} \max_{(e,d) \in \overline{D}_x} \min\{h_e(e_\gamma), h_d(d_\delta)\}$$

where $\overline{D}_x$ is the set of all pairs (e_γ, d_δ) satisfying the property (3.50).
4) Prove if

$$\Phi_1(u_i;e_\gamma) \geq y_{min} \qquad \text{and} \qquad \Phi_2(u_i;d_\delta) \leq y_{max} \qquad (3.51)$$

for

$$\gamma = 1, 2, ..., v-1, v+1, ..., s,$$

$$\delta = 1, 2, ..., \mu-1, \mu+1, ..., l.$$

5) Determine

$$v_{ci} = \frac{1}{2} \max_{(c,d) \in D_x} \min \{h_e(e_\gamma), h_d(d_\delta)\}$$

where D_x is the set of all pairs (e_γ, d_δ) satisfying the property (3.51).

6) Execute the points $1-4$ for $i = 1, 2, ..., p$.

7) Choose

$$i^* = \arg \max_{i \in \overline{1,p}} v_{ci}.$$

The result (the optimal decision) is: $\hat{u} = u_i$ for $i = i^*$.

The algorithm is illustrated in Fig. 3.9. For the great size of the problem (the great value p) the method of *integer programming* may be used to determine i^*.

Example 3.7: One-dimensional plant is described by the inequality

$$xu \leq y \leq 2xu,$$

$$u \in \{1, 2, 3\}, \qquad x \in \{3, 4, 5, 6\}$$

and the corresponding values of $h_x(x)$ are $(0.5, 0.6, 1, 0.4)$. The requirement is $y \in D_y = [5, 10]$. Then $D_x(D_y, u)$ is determined by

$$\frac{5}{u} \leq x \leq \frac{10}{u}.$$

For $u = 1$

$$v_1 = v\{\overline{x} \tilde{\in} [5, 10]\} = \max\{h_x(5), h_x(6)\} = 1.$$

For $u = 2$

$$v_2 = v\{\overline{x} \tilde{\in} [2.5, 5]\} = \max\{h_x(3), h_x(4), h_x(5)\} = 1.$$

For $u = 3$

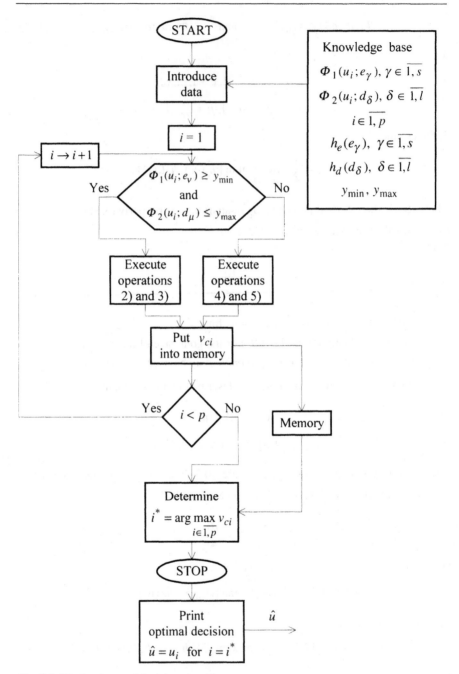

Fig. 3.9. Block scheme of decision algorithm

$$v_3 = v\{\bar{x} \tilde{\in} [\frac{5}{3}, \frac{10}{3}]\} = h_x(3) = 0.5.$$

Then $\hat{u} = 1$ or 2, and $v(\hat{u}) = 1$.

Now let us assume that $\bar{x}$ is C-uncertain variable.

For $u = 1$

$$v_{c1} = v_c\{\bar{x} \tilde{\in} [5, 10]\} = \frac{1}{2}[1 + 1 - \max\{h_x(3), h_x(4)\}] = 0.7.$$

For $u = 2$

$$v_{c2} = v_c\{\bar{x} \tilde{\in} [2.5, 5]\} = \frac{1}{2}[1 + 1 - h_x(6)] = 0.8.$$

For $u = 3$

$$v_{c3} = v_c\{\bar{x} \tilde{\in} [\frac{5}{3}, \frac{10}{3}]\} = \frac{1}{2}[0.5 + 1 - \max\{h_x(4), h_x(5), h_x(6)\}] = 0.25.$$

Then $\hat{u}_c = 2$ and $v_c(\hat{u}_c) = 0.8$, i.e. for $u = 2$ the certainty index that the set of possible outputs belongs to the set $[5, 10]$ is equal to 0.8.

4 Systems with Logical Knowledge Representation

4.1 Logical Knowledge Representation

Now we shall consider the knowledge representation in which the relations R_i (3.3) have the form of logic formulas concerning u, y, w. Let us introduce the following notation:

1. $\alpha_{uj}(u)$ – simple formula (i.e. simple property) concerning u, $j = 1, 2, ..., n_1$,

   e.g. $\alpha_{u1}(u) = "u^T u \leq 2"$.

2. $\alpha_{wr}(u, w, y)$ – simple formula concerning u, w and y, $r = 1, 2, ..., n_2$.

3. $\alpha_{ys}(y)$ – simple formula concerning y, $s = 1, 2, ..., n_3$.

4. $\alpha_u = (\alpha_{u1}, \alpha_{u2}, ..., \alpha_{un_1})$ – subsequence of simple formulas concerning u.

5. $\alpha_w = (\alpha_{w1}, \alpha_{w2}, ..., \alpha_{wn_2})$ – subsequence of simple formulas concerning u, w and y.

6. $\alpha_y = (\alpha_{y1}, \alpha_{y2}, ..., \alpha_{yn_3})$ – subsequence of simple formulas concerning y.

7. $\alpha(u, w, y) \triangleq (\alpha_1, \alpha_2, ..., \alpha_n) = (\alpha_u, \alpha_w, \alpha_y)$ – sequence of all simple formulas in the knowledge representation, $n = n_1 + n_2 + n_3$.

8. $F_i(\alpha)$ – the i-th fact given by an expert. It is a logic formula composed of the subsequence of α and the logic operations: $\vee$ – *or*, $\wedge$ – *and*, $\neg$ – *not*, $\rightarrow$ – *if ... then*, $i = 1, 2, ..., k$.

   For example $F_1 = \alpha_1 \wedge \alpha_2 \rightarrow \alpha_4$, $F_2 = \alpha_3 \vee \alpha_2$ where $\alpha_1 = "u^T u \leq 2"$, $\alpha_2 = "$ the temperature is small or $y^T y \leq 3"$, $\alpha_3 = "y^T y > w^T w"$, $\alpha_4 = "y^T y = 4"$.

9. $F(\alpha) = F_1(\alpha) \wedge F_2(\alpha) \wedge ... \wedge F_k(\alpha)$.

10. $F_u(\alpha_u)$ – input property, i.e. the logic formula using α_u.

11. $F_y(\alpha_y)$ – output property.

12. $a_m \in \{0,1\}$ – logic value of the simple property α_m, $m = 1, 2, ..., n$.

13. $a = (a_1, a_2, ..., a_n)$ – zero-one sequence of the logic values.

14. $a_u(u)$, $a_w(u, w, y)$, $a_y(y)$ – zero-one subsequences of the logic values corresponding to $\alpha_u(u)$, $\alpha_w(u, w, y)$, $\alpha_y(y)$.

15. $F(a)$ – the logic value of $F(\alpha)$.

All facts given by an expert are assumed to be true, i.e. $F(a) = 1$.

The description

$$< \alpha, F(\alpha) > \overset{\Delta}{=} KP$$

may be called a *logical knowledge representation* of the plant. For illustration purposes let us consider a very simple example:

$u = (u^{(1)}, u^{(2)})$, $y = (y^{(1)}, y^{(2)})$, $w \in R^1$,

$\alpha_{u1} = " u^{(1)} + u^{(2)} > 0"$, $\alpha_{u2} = " u^{(2)} > 2"$, $\alpha_{y1} = " y^{(2)} < y^{(1)} "$,

$\alpha_{y2} = " y^{(1)} + y^{(2)} = 4"$, $\alpha_{w1} = " u^{(1)} - 2w + y^{(2)} < 0"$,

$\alpha_{w2} = " u^{(2)} > y^{(1)} "$,

$F_1 = \alpha_{u1} \wedge \alpha_{w1} \rightarrow \alpha_{y1} \vee \neg \alpha_{w2}$, $F_2 = (\alpha_{u2} \wedge \alpha_{w2}) \vee (\alpha_{y2} \wedge \neg \alpha_{u1})$,

$F_u = \alpha_{u1} \vee \alpha_{u2}$, $F_y = \neg \alpha_{y2}$.

The expressions $F(a)$ have the same form as the formulas $F(\alpha)$, e.g.

$$F_1(a_{u1}, a_{w1}, a_{w2}, a_{y1}) = a_{u1} \wedge a_{w1} \rightarrow a_{y1} \vee \neg a_{w2}.$$

The logic formulas $F_j(\alpha)$, $F_u(\alpha_u)$ and $F_y(\alpha_y)$ are special forms of the relations introduced in Sects. 3.1 and 3.2. Now the relation (3.3) has the form

$$R_i(u, w, y) = \{(u, w, y) \in U \times W \times Y : F_i[a(u, w, y)] = 1\}, \quad i \in \overline{1, k}. \quad (4.1)$$

The input and output properties may be expressed as follows:

$$u \in D_u, \qquad y \in D_y$$

where

$$D_u = \{u \in U : F_u[a_u(u)] = 1\}, \qquad (4.2)$$

$$D_y = \{y \in Y : F_y[a_y(y)] = 1\}. \qquad (4.3)$$

The description with $F(a)$, $F_u(a_u)$, $F_y(a_y)$ may be called the *description on the logical level*. The expressions $F(a)$, $F_u(a_u)$ and $F_y(a_y)$ describe *logical structures* of the plant, the input property and the output property, respectively. The description on the logical level is independent of the particular meaning of the simple formulas. In other words, it is common for the different plants with differ-ent practical descriptions but the same logical structures. On the logical level our plant may be considered as a relational plant with the input a_u (a vector with n_1 zero-one components) and the output a_y (a vector with n_3 zero-one compo-nents), described by the relation

$$F(a_u, a_w, a_y) = 1 \qquad (4.4)$$

(Fig. 4.1). The input and output properties for this plant corresponding to the properties $u \in D_u$ and $y \in D_y$ for the plant with input u and output y are as follows

$$a_u \in \bar{S}_u \subset S_u, \qquad a_y \in \bar{S}_y \subset S_y$$

where S_u, S_y are the sets of all zero-one sequences a_u, a_y, respectively, and

$$\bar{S}_u = \{a_u \in S_u : F_u(a_u) = 1\}, \qquad \bar{S}_y = \{a_y \in S_y : F_y(a_y) = 1\}. \quad (4.5)$$

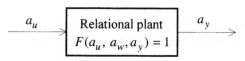

Fig. 4.1. Plant on logical level

4.2 Analysis and Decision Making Problems

The analysis and decision making problems for the relational plant described by the logical knowledge representation are analogous to those for the relational plant in Sect. 3.2. The analysis problem consists in finding the output property for the given input property and the decision problem is an inverse problem consisting in finding the input property (the decision) for the required output property.

Analysis problem: For the given $F(\alpha)$ and $F_u(\alpha_u)$ find the best property $F_y(\alpha_y)$ such that the implication

$$F_u(\alpha_u) \rightarrow F_y(\alpha_y) \qquad (4.6)$$

is satisfied.

If it is satisfied for F_{y1} and F_{y2}, and $F_{y1} \rightarrow F_{y2}$, then F_{y1} is better than F_{y2}. The property F_y is then the best if it implies any other property for which the implication (4.6) is satisfied. The best property F_y corresponds to the smallest set D_y in the formulation presented in Sect. 3.2.

Decision problem: For the given $F(\alpha)$ and $F_y(\alpha_y)$ (the property required by a user) find the best property $F_u(\alpha_u)$ such that the implication (4.6) is satisfied.

If it is satisfied for F_{u1} and F_{u2}, and $F_{u2} \rightarrow F_{u1}$, then F_{u1} is better than F_{u2}. The property F_u is then the best if it is implied by any other property for which the implication (4.6) is satisfied. The best property F_u corresponds to the largest set D_u in the formulation presented in Sect. 3.2.

Remark 4.1: The solution of our problem may not exist. In the case of the analysis it means that there is a contradiction between the property $F_u(\alpha_u)$ and the facts $F(\alpha_u, \alpha_w, \alpha_y)$, i.e. the sequence a_u such that $F_u(a_u) \wedge F(a_u, a_w, a_y) = 1$ does not exist. In the case of the decision making it means that the requirement F_y is too strong. The existence of the solution will be explained in the next section.

Remark 4.2: Our problems are formulated and will be solved on the logic level. Consequently they depend on the logical structures (the form of F and F_y or F_u) but do not depend on the meaning of the simple formulas. The knowledge representation KP and the problem formulations may be extended for different variables, objects and sets (not particularly the sets of real number vectors) used in the description of the knowledge. For example, in the example in the previous section we may have the following simple formulas in the text given by an expert:

α_{u1} = "operation O_1 is executed after operation O_2",

α_{u2} = "temperature is small",

α_{w1} = "pressure is high",

α_{w2} = "humidity is small",

α_{y1} = "state S occurs",

α_{y2} = "quality of product is sufficient".

Then the facts F_1 and F_2 in this example mean:

F_1 = "If operation O_1 is executed after operation O_2 and pressure is high then state S occurs or humidity is not small",

F_2 = "Temperature is small and humidity is small or quality is sufficient and operation O_1 is not executed after operation O_2".

Remark 4.3: The possibilities of forming the input and output properties are restricted. Now the sets D_u and D_y may be determined by the logic formulas

$F_u(\alpha_u)$ and $F_y(\alpha_y)$ using the simple formulas α_u and α_y from the sequence of the simple formulas α used in the knowledge representation.

4.3 Logic-Algebraic Method

The solutions of the analysis and decision problems formulated in Sect. 4.2 may be obtained by using so called *logic-algebraic method* [9, 13, 14, 19]. It is easy to show that the analysis problem is reduced to solving the following algebraic equation

$$\tilde{F}(a_u, a_w, a_y) = 1 \qquad (4.7)$$

with respect to a_y, where

$$\tilde{F}(a_u, a_w, a_y) = F_u(a_u) \wedge F(a_u, a_w, a_y).$$

Now $F(a_u, a_w, a_y)$, $F_u(a_u)$ and $F_y(a_y)$ are algebraic expressions in two-value logic algebra. If S_y is the set of all solutions then F_y is determined by S_y, i.e. $a_y \in S_y \leftrightarrow F_y(a_y) = 1$. For example, if $a_y = (a_{y1}, a_{y2}, a_{y3})$ and $S_y = \{(1,1,0),(0,1,0)\}$ then $F_y(\alpha_y) = (\alpha_{y1} \wedge \alpha_{y2} \wedge \neg \alpha_{y3}) \vee (\neg \alpha_{y1} \wedge \alpha_{y2} \wedge \neg \alpha_{y3})$.

In the decision making problem two sets of the algebraic equations should be solved with respect to a_u :

$$\left.\begin{array}{r}F(a_u, a_w, a_y) = 1 \\ F_y(a_y) = 1\end{array}\right\}, \qquad \left.\begin{array}{r}F(a_u, a_w, a_y) = 1 \\ F_y(a_y) = 0\end{array}\right\} \qquad (4.8)$$

If S_{u1}, S_{u2} are the sets of the solutions of the first and the second equation, respectively – then $F_u(\alpha_u)$ is determined by $S_u = S_{u1} - S_{u2}$ [13] in the same way as F_y by S_y in the former problem.

The generation of the set S_y requires the testing of all sequences $a = (a_u, a_w, a_y)$ and the execution time may be very long for the large size of the problem. The similar computational difficulties may be connected with the solution of the decision problem. The generation of S_y (and consequently, the solution F_y) may be much easier when the following decomposition is applied:

$$F_u \wedge F = \overline{F}_1(\bar{a}_0, \bar{a}_1) \wedge \overline{F}_2(\bar{a}_1, \bar{a}_2) \wedge \ldots \wedge \overline{F}_N(\bar{a}_{N-1}, \bar{a}_N) \qquad (4.9)$$

where $\bar{a}_0 = a_y$, $\overline{F}_1$ is the conjunction of all facts from $\tilde{F}$ containing the variables from $\bar{a}_0$, $\bar{a}_1$ is the sequence of all other variables in $\overline{F}_1$, $\overline{F}_2$ is the conjunction of all facts containing the variables from $\bar{a}_1$, $\bar{a}_2$ is the sequence of all other variables in $\overline{F}_2$ etc. As a result of the decomposition the following *recursive procedure* may be applied to obtain $\overline{S}_0 = S_y$:

$$\overline{S}_{m-1} = \{\bar{a}_{m-1} \in S_{m-1} : \bigvee_{\bar{a}_m \in \overline{S}_m} [\overline{F}_m(\bar{a}_{m-1}, \bar{a}_m) = 1]\}, \qquad (4.10)$$

where S_m is the set of all $\bar{a}_m$, $m = N, N-1, \ldots, 1$, $\overline{S}_N = S_N$.

The recursive procedure (4.10) has two interesting interpretations:

A. System analysis interpretation.

Let us consider the cascade of relation elements (Fig. 4.2) with input $\bar{a}_m$, output $\bar{a}_{m-1}$ (zero-one sequences), described by the relations $\overline{F}_m(\bar{a}_{m-1}, \bar{a}_m) = 1$ ($m = N, N-1, \ldots, 1$). Then $\overline{S}_{m-1}$ is the set of all possible outputs from the element $\overline{F}_m$ and $\overline{S}_0$ is the set of all possible outputs from the whole cascade.

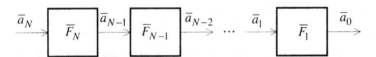

Fig. 4.2. Relational system

B. Deductive reasoning interpretation.

The set $\overline{S}_{m-1}$ may be considered as the set of all elementary conclusions from $\overline{F}_N \wedge \ldots \wedge \overline{F}_m$, and $\overline{S}_0$ is the set of all elementary conclusions from the facts $F_u \wedge F$.

A similar approach may be applied to the decision problem. To determine S_{y1} and S_{y2} we may use the recursive procedure (4.10) with F in (4.9) instead of $F_u \wedge F$ and with $\bar{a}_0 = (a_u, a_y)$. After the generation of $\overline{S}_0$ from (4.10) one can determine S_{u1} and S_{u2} in the following way:

$$S_{u1} = \{a_u : \bigvee_{a_y \in \overline{S}_y} [(a_u, a_y) \in \overline{S}_0]\},$$

$$S_{u2} = \{a_u : \bigvee_{a_y \in \hat{S}_y - \bar{S}_y} [(a_u, a_y) \in \bar{S}_0]\}$$

where $\bar{S}_y = \{a_y : F_y(a_y) = 1\}$ and $\hat{S}_y$ is the set of all a_y.

The different versions of the presented procedures have been elaborated and applied in the general purpose expert systems CONTROL-LOG and CLASS-LOG, specially oriented for the applications to a class of knowledge-based control systems and to classification problems.

The main idea of the logic-algebraic method presented here for the generation of the solutions consists in replacing the individual reasoning concepts based on inference rules by unified algebraic procedures based on the rules in two-value logic algebra. The results may be considered as a unification and generalization of the different particular reasoning algorithms (see e.g. [3]) for a class of the systems with the logical knowledge representation for which the logic-algebraic method has been developed. The logic-algebraic method can be applied to the design of complex knowledge-based computer systems [43, 45, 46, 48].

Example 4.1 (analysis): The facts $\tilde{F}$ are the following:

$F_1 = (\alpha_3 \vee \neg \alpha_1) \rightarrow \alpha_4$, $F_2 = (\neg \alpha_1 \wedge \alpha_7) \vee \neg \alpha_3$, $F_3 = (\alpha_9 \wedge \alpha_1) \rightarrow \alpha_2$,

$F_4 = (\alpha_4 \wedge \neg \alpha_7) \vee \alpha_5$, $F_5 = \alpha_6 \rightarrow (\alpha_4 \wedge \alpha_8)$, $F_6 = \alpha_2 \rightarrow (\neg \alpha_4 \wedge \alpha_6)$,

$F_7 = (\alpha_3 \wedge \alpha_2) \vee \alpha_{10}$, $\alpha_y = (\alpha_9, \alpha_{10})$.

It is not important which simple formulas from $\alpha_1 \div \alpha_8$ are α_u and which fact from the set $\{ F_1, F_2, F_4, F_5, F_6 \}$ (not containing α_y) is the input property. It is easy to see that

$\bar{F}_1(\bar{a}_0, \bar{a}_1) = F_3(a_1, a_2, a_9) \wedge F_7(a_2, a_3, a_{10})$, $\bar{a}_1 = (a_1, a_2, a_3)$,

$\bar{F}_2(\bar{a}_1, \bar{a}_2) = F_1(a_1, a_3, a_4) \wedge F_2(a_1, a_3, a_7) \wedge F_6(a_2, a_4, a_6)$,

$\bar{a}_2 = (a_4, a_6, a_7)$,

$\bar{F}_3(\bar{a}_2, \bar{a}_3) = F_4(a_4, a_5, a_7) \wedge F_5(a_4, a_6, a_8)$, $\bar{a}_3 = (a_5, a_8)$.

In our case $N = 3$, $S_N = \{(1, 1), (1, 0), (0, 1), (0, 0)\}$. According to (4.10) one should put successively the elements of S_N into $\bar{F}_3$ and determine all 0-1 sequences (a_4, a_6, a_7) such that $\bar{F}_3 = 1$. These are the elements of $\bar{S}_2$. In a similar way one determines $\bar{S}_1$ and finally $\bar{S}_0 = \{(0, 1), (1, 1)\}$. Then $F_y = (\neg \alpha_9 \wedge \alpha_{10}) \vee (\alpha_9 \wedge \alpha_{10}) = \alpha_{10}$.

Example 4.2 (decision making): The facts F in the knowledge representation KP are the following:

$F_1 = \alpha_1 \wedge (\alpha_4 \vee \neg\alpha_6)$, $F_2 = (\alpha_2 \wedge \alpha_4) \rightarrow \alpha_6$, $F_3 = \neg\alpha_4 \vee \neg\alpha_3 \vee \alpha_5$, $F_4 = \alpha_4 \wedge (\alpha_3 \vee \neg\alpha_5)$, $F_5 = (\alpha_4 \wedge \neg\alpha_2) \rightarrow \alpha_7$, $\alpha_u = (\alpha_1, \alpha_2)$, $\alpha_y = (\alpha_6, \alpha_7)$.

Now $\bar{a}_0 = (a_u, a_y) = (a_1, a_2, a_6, a_7)$, $\bar{F}_1 = F_1 \wedge F_2 \wedge F_5$, $\bar{F}_2 = F_3 \wedge F_4$, $\bar{a}_1 = a_4$, $\bar{a}_2 = (a_3, a_5)$.

Using (4.10) (two steps for $m = 2, 1$) we obtain $\bar{S}_0 = \{(1,1,1,1), (1,1,1,0), (1,0,1,1), (1,0,0,1)\}$. We can consider the different cases of $F_y(\alpha_6, \alpha_7)$. It is easy to see that for $F_y = \alpha_6 \vee \alpha_7$ we have $S_y = \{(1,1), (1,0), (0,1)\}$, $S_{u1} = \{(1,1), (1,0)\}$, S_{u2} is an empty set, $S_u = S_{u1}$ and $F_u = (\alpha_1 \wedge \alpha_2) \vee (\alpha_1 \wedge \neg\alpha_2) = \alpha_1$. If $F_y = \alpha_6$ then $F_u = \alpha_1 \wedge \alpha_2$, if $F_y = \alpha_7$ then $F_u = \alpha_1 \wedge \neg\alpha_2$, if $F_y = \alpha_6 \wedge \alpha_7$ then $S_{u1} = S_{u2}$, S_u is an empty set and the solution F_u does not exist.

The formulas α and the facts may have a different practical sense. For example, in the second example $u, y, c \in R^1$ and: $\alpha_1 = "u \leq 3c"$, $\alpha_2 = "u^2 + c^2 \leq 1"$, $\alpha_3 = "$pressure is high$"$, $\alpha_4 = "$humidity is small$"$, $\alpha_5 = "$temperature is less than $u + y + c$ $"$, $\alpha_6 = "y^2 + (c - 0.5)^2 \leq 0.25"$, $\alpha_7 = "-c \leq y \leq c"$ for a given parameter c. For example, the fact F_2 means that: $"$if $u^2 + c^2 \leq 1$ and humidity is small then $y^2 + (c - 0.5)^2 \leq 0.25"$, the fact F_3 means that: $"$humidity is not small or pressure is not high or temperature is less than $u + y + c"$. The required output property $F_y = \alpha_6$ is obtained if $F_u = \alpha_1 \wedge \alpha_2$, i.e. if $u \leq 3c$ and $u^2 + c^2 \leq 1$.

4.4 Analysis and Decision Making for the Plant with Uncertain Parameters

Now let us consider the plant described by a logical knowledge representation with uncertain parameters in the simple formulas and consequently in the properties F, F_u, F_y [11, 26]. In general, we may have the simple formulas $\alpha_u(u; x)$, $\alpha_w(u, y, w; x)$ and $\alpha_y(y; x)$ where $x \in X$ is an unknown vector parameter which is assumed to be a value of an uncertain variable $\bar{x}$ with the certainty distribution $h_x(x)$ given by an expert. For example,

$$\alpha_{u1} = "u^T u \leq 2x^T x", \quad \alpha_{w1} = "y^T y \leq x^T x", \quad \alpha_{y1} = "y^T y + x^T x < 4".$$

In particular, only some simple formulas may depend on some components of the vector x.

In the analysis problem the formula $F_u[\alpha_u(u;x)]$ depending on x means that the observed (given) input property is formulated with the help of the unknown parameter (e.g. we may know that u is less than the temperature of a raw material x, but we do not know the exact value of x). Solving the analysis problem described in Sects. 4.2 and 4.3 we obtain $F_y[\alpha_y(y;x)]$ and consequently

$$D_y(x) = \{ y \in Y : \ F_y[a_y(y;x)] = 1 \} .$$

Further considerations are the same as in Sect. 3.4 for the given set D_u. In version II (see (3.33) and (3.34)) we have

$$v[D_y(\bar{x}) \subseteq D_y] = \max_{x \in D_x(D_y)} h_x(x)$$

where D_y is given by a user and

$$D_x(D_y) = \{ x \in X : \ D_y(x) \subseteq D_y \} .$$

In the decision problem the formula $F_y[\alpha_y(y;x)]$ depending on x means that the user formulates the required output property with the help of the unknown parameter (e.g. he wants to obtain y less than the temperature of a product x). Solving the decision problem described in Sects. 4.2 and 4.3 we obtain $F_u[(u;x)]$ and consequently

$$D_u(x) = \{ u \in U : \ F_u[a_u(u;x)] = 1 \} . \tag{4.11}$$

Further considerations are the same as in Sect. 3.5 for version II (see (3.44)). The optimal decision, maximizing the certainty index that the requirement $F_y[\alpha_y(y;x)]$ is satisfied, may be obtained in the following way:

$$\hat{u} = \arg\max_u \ \max_{x \in D_{xd}(u)} h_x(x)$$

where

$$D_{xd}(u) = \{ x \in X : \ u \in D_u(x) \} .$$

Example 4.3:   The facts are the same as in example 4.2 where $c \triangleq x$. In the example 4.2 for the required output property $F_y = \alpha_6$ the following result has been obtained:

If

$$u \leq 3x \qquad \text{and} \qquad u^2 + x^2 \leq 1 \qquad (4.12)$$

then

$$y^2 + (x - 0.5)^2 \leq 0.25 .$$

The inequalities (4.12) determine the set (4.11) in our case. Assume that x is a value of an uncertain variable with triangular certainty distribution: $h_x = 2x$ for $0 \leq x \leq \frac{1}{2}$, $h_x = -2x + 2$ for $\frac{1}{2} \leq x \leq 2$, $h_x = 0$ otherwise. Then we can use the result in Example 3.5. As the decision $\hat{u}$ we can choose any value from $[-\frac{\sqrt{3}}{2}, \frac{\sqrt{3}}{2}]$ and the requirement will be satisfied with the certainty index equal to 1. The result for C-uncertain variable is $\hat{u}_c = 0$ and $v_c(\hat{u}_c) = 1$.

4.5   Uncertain Logical Decision Algorithm

Consider the plant with external disturbances $z \in Z$. Then in the logical knowledge representation we have the simple formulas $\alpha_u(u, z; x)$, $\alpha_w(u, w, y, z; x)$, $\alpha_y(y, z; x)$ and $\alpha_z(z; x)$ to form the property $F_z(\alpha_z)$ concerning z.

The analysis problem analogous to that described in Sects. 4.2 and 4.3 for the fixed x is as follows: For the given $F(\alpha_u, \alpha_w, \alpha_y, \alpha_z)$, $F_z(\alpha_z)$ and $F_u(\alpha_u)$ find the best property $F_y(\alpha_y)$ such that the implication

$$F_z(\alpha_z) \wedge F_u(\alpha_u) \rightarrow F_y(\alpha_y) \qquad (4.13)$$

is satisfied. In this formulation $F_z(\alpha_z)$ denotes an observed property concerning z.

The problem solution is the same as in Sect. 4.3 with $F_z \wedge F_u$ in the place of F_u. As a result one obtains $F_y[\alpha_y(y, z; x)]$ and consequently

$$D_y(z; x) = \{y \in Y : \ F_y[a_y(y, z; x)] = 1\} .$$

Further considerations are the same as in Sect. 3.4.

The *decision problem* analogous to that described in Sects. 4.2 and 4.3 for the fixed x is as follows: For the given $F(\alpha_u, \alpha_w, \alpha_y, \alpha_z)$, $F_z(\alpha_z)$ and $F_y(\alpha_y)$ find the best property $F_u(\alpha_u)$ such that the implication (4.13) is satisfied.

The problem solution is the same as in Sect. 4.3 with $F_z \wedge F$ in the place of F. As a result one obtains

$$D_u(z\,;x) = \{\, u \in U \,:\; F_u[a_y(u,z\,;x)] = 1 \,\}. \qquad (4.14)$$

Further considerations are the same as in Sect. 3.5 for version II.

To obtain the solution of the decision problem another approach may be applied. For the given F and F_y we may state the problem of finding the best input property $F_d(\alpha_u,\alpha_z)$ such that the implication

$$F_d(\alpha_u,\alpha_z) \to F_y(\alpha_y)$$

is satisfied. The solution may be obtained in the same way as in Sect. 4.3 with (α_u,α_z) and F_d in the place of α_u and F_u, respectively. The formula $F_d(\alpha_u,\alpha_z)$ may be called a logical knowledge representation for the decision making (i.e. the logical form of KD) or a *logical uncertain decision algorithm* corresponding to the relation $\bar{R}$ or the set $D_u(z\,;x)$ in Sect. 3.5. For the given $F_z(\alpha_z)$, the input property may be obtained in the following way: Denote by S the set of all (a_u,a_z) for which $F_d = 1$ and by $\bar{S}_z$ the set of all a_z for which $F_z = 1$, i.e.

$$\bar{S}_d = \{(a_u,a_z): \; F_d(a_u,a_z) = 1\}.$$

$$\bar{S}_z = \{a_z: \; F_z(a_z) = 1\}.$$

Then $F_u(\alpha_u)$ is determined by the set

$$S_u = \{\, a_u \in S_u \,:\; \bigwedge_{a_z \in \bar{S}_z} (a_u,a_z) \in \bar{S}_d \,\}. \qquad (4.15)$$

The formula (4.14) is analogous to the formula (3.16) for the relational plant. It follows from the fact that on the logical level our plant may be considered as a relational plant with the input a_u, the disturbance a_z and the output a_y (see Fig. 4.1).

5 Dynamical Systems

5.1 Relational Knowledge Representation [12]

The relational knowledge representation for the dynamical plant may have the form analogous to that for the static plant presented in Sect. 3.1. The deterministic dynamical plant is described by the equations

$$\left.\begin{array}{l} s_{n+1} = f(s_n, u_n), \\ y_n = \eta(s_n) \end{array}\right\} \tag{5.1}$$

where n denotes the discrete time and $s_n \in S$, $u_n \in U$, $y_n \in Y$ are the state, the input and the output vectors, respectively. In the relational dynamical plants the functions f and η are replaced by relations

$$\left.\begin{array}{l} R_I(u_n, s_n, s_{n+1}) \subseteq U \times S \times S, \\ R_{II}(s_n, y_n) \subseteq S \times Y. \end{array}\right\} \tag{5.2}$$

The relations R_I and R_{II} form a *relational knowledge representation of the dynamical plant*. For a nonstationary plant the relations R_I and R_{II} depend on n. The relations R_I and R_{II} may have the form of equalities and/or inequalities concerning the components of the respective vectors. In particular the relations are described by inequalities

$$f_1(u_n, s_n) \leq s_{n+1} \leq f_2(s_n, u_n),$$
$$\eta_1(s_n) \leq y_n \leq \eta_2(s_n),$$

i.e. by a set of inequalities for the respective components of the vectors. The formulations of the analysis and decision problems may be similar to those in Sect. 3.2. Let us assume that $s_0 \in D_{s0} \subset S$.

Analysis problem: For the given relations (5.2), the set D_{s0} and the given sequence of sets $D_{un} \subset U$ ($n = 0, 1, ...$) one should find a sequence of the smallest sets $D_{yn} \subset Y$ ($n = 1, 2, ...$) for which the implication

$$(u_0 \in D_{u0}) \wedge (u_1 \in D_{u1}) \wedge \dots \wedge (u_{n-1} \in D_{u,n-1}) \to y_n \in D_{yn}$$

is satisfied.

It is an extension of the analysis problem for the deterministic plant (5.1), consisting in finding the sequence y_n for the given sequence u_n and the initial state s_0, and for the known functions f, η. For the fixed moment n our plant may be considered as a connection of two static relational plants (Fig. 5.1). The analysis problem is then reduced to the analysis for the relational plants R_I and R_{II}, described in Sect. 3.2. Consequently, according to the formula (3.5) applying to R_I and R_{II}, we obtain the following *recursive procedure* for $n = 1, 2, \dots$:

1. For the given D_{un} and D_{sn} obtained in the former step, determine the set $D_{s,n+1}$ using $R_I(u_n, s_n, s_{n+1})$:

$$D_{s,n+1} = \{s_{n+1} \in S : \bigvee_{u_n \in D_{un}} \bigvee_{s_n \in D_{sn}} [(u_n, s_n, s_{n+1}) \in R_I(u_n, s_n, s_{n+1})]\}. \quad (5.3)$$

2. Using $D_{s,n+1}$ and $R_{II}(s_{n+1}, y_{n+1})$, determine $D_{y,n+1}$:

$$D_{y,n+1} = \{y_{n+1} \in Y : \bigvee_{s_{n+1} \in D_{s,n+1}} [(s_{n+1}, y_{n+1}) \in R_{II}(s_{n+1}, y_{n+1})]\}. \quad (5.4)$$

For $n = 0$ in the formula (5.3) we use the given set D_{s0}.

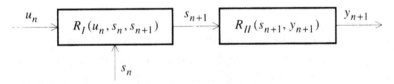

Fig. 5.1. Dynamical relational plant

Decision problem: For the given relations (5.2), the set D_{s0} and the sequence of sets $D_{yn} \subset Y$ ($n = 1, 2, \dots, N$) one should determine the sequence D_{un} ($n = 0, 1, \dots, N-1$) such that the implication

$$(u_0 \in D_{u0}) \wedge (u_1 \in D_{u1}) \wedge \dots \wedge (u_{N-1} \in D_{u,N-1})$$
$$\to (y_1 \in D_{y1}) \wedge (y_2 \in D_{y2}) \wedge \dots \wedge (y_N \in D_{y,N})$$

is satisfied.

The set D_{yn} is given by a user and the property $y_n \in D_{yn}$ $(n = 1, 2, ..., N)$ denotes the user's requirement. To obtain the solution one can apply the following *recursive procedure* starting from $n = 0$:

1. For the given $D_{y,n+1}$, using $R_{II}(s_{n+1}, y_{n+1})$ determine the largest set $D_{s,n+1}$ for which the implication

$$s_{n+1} \in D_{s,n+1} \rightarrow y_{n+1} \in D_{y,n+1}$$

is satisfied. This is a decision problem for the part of the plant described by R_{II} (see Fig. 5.1). According to (3.13) with s_{n+1}, y_{n+1} in the place of (u, y) we obtain

$$D_{s,n+1} = \{s_{n+1} \in S : D_{y,n+1}(s_{n+1}) \subseteq D_{y,n+1}\} \tag{5.3a}$$

where

$$D_{y,n+1}(s_{n+1}) = \{y_{n+1} \in Y : (s_{n+1}, y_{n+1}) \in R_{II}(s_{n+1}, y_{n+1})\}.$$

2. For $D_{s,n+1}$ obtained at the point 1 and D_{sn} obtained in the former step, using $R_I(u_n, s_n, s_{n+1})$ determine the largest set D_{un} for which the implication

$$(u_n \in D_{un}) \wedge (s_n \in D_{sn}) \rightarrow s_{n+1} \in D_{s,n+1}$$

is satisfied. This is a decision problem for the part of the plant described by R_I. According to (3.16) with (u_n, s_{n+1}, s_n) in the place of (u, y, z) we obtain

$$D_{un} = \{u_n \in U : \bigwedge_{s_n \in D_{sn}} [D_{s,n+1}(u_n, s_n) \subseteq D_{s,n+1}]\} \tag{5.4a}$$

where

$$D_{s,n+1}(u_n, s_n) = \{s_{n+1} \in S : (u_n, s_n, s_{n+1}) \in R_I(u_n, s_n, s_{n+1})\}.$$

Remark 5.1: In the formulation of the decision problem we did not use the statement "the largest set D_{un}". Now the set of all possible decisions means the set of all sequences $u_0, u_1, ..., u_{N-1}$ for which the requirements are satisfied. Using the recursive procedure described above we do not obtain the set of all possible decisions. In other words, we determine the set of sequences $u_0, u_1, ..., u_{N-1}$ belonging to the set of all input sequences for which the requirements concerning y_n are satisfied.

Remark 5.2: The relations R_I and R_{II} may be given by the sets of facts in a similar way as described in Sect. 4.1. The formulation and solution of the analysis and decision problems for the plant described by *dynamical logical knowledge representation* are analogous to those presented above and for the fixed n are reduced to the analysis and decision problems considered for the static plant in Sect. 4.2.

The considerations presented in this section will be used for the plants with uncertain parameters in the knowledge representation, described in the next section.

Example 5.1:   As a very simple example let us consider first order one-dimensional plant described by inequalities

$$a_1 s_n + b_1 u_n \leq s_{n+1} \leq a_2 s_n + b_2 u_n,$$

$$c_1 s_{n+1} \leq y_{n+1} \leq c_2 s_{n+1}.$$

It is known that $s_{01} \leq s_0 \leq s_{02}$; b_1, b_2, c_1, $c_2 > 0$. The requirement concerning y_n is as follows

$$\bigwedge_{n \geq 1} (y_{\min} \leq y_n \leq y_{\max}),$$

i.e. $D_{yn} = [y_{\min}, y_{\max}]$ for every n. For the given s_{01}, s_{02}, $y_{\min}$, $y_{\max}$ and the coefficients a_1, a_2, b_1, b_2, c_1, c_2 one should determine the sequence D_{un} such that if $u_n \in D_{un}$ for every n then the requirement is satisfied. For $n = 0$ the set D_{s1} according to (5.3a) is determined by inequalities

$$c_2 s_1 \leq y_{\max}, \qquad c_1 s_1 \geq y_{\min}.$$

Then

$$D_{s1} = [\frac{y_{\min}}{c_1}, \frac{y_{\max}}{c_2}].$$

Using (5.4a) for u_0 we obtain the following inequalities

$$a_2 s_{02} + b_2 u_0 \leq \frac{y_{\max}}{c_2},$$

$$a_1 s_{01} + b_1 u_0 \geq \frac{y_{\min}}{c_1}$$

and

$$D_{u0} = [\frac{y_{\min}}{b_1 c_1} - \frac{a_1 s_{01}}{b_1}, \frac{y_{\max}}{b_2 c_2} - \frac{a_2 s_{02}}{b_2}].$$

For $n \geq 1$ $D_{s,n+1} = D_{s1}$ and according to (5.4a) D_{un} is determined by inequalities

$$a_2 \frac{y_{\max}}{c_2} + b_2 u_n \leq \frac{y_{\max}}{c_2},$$

$$a_1 \frac{y_{\min}}{c_1} + b_1 u_n \geq \frac{y_{\min}}{c_1}.$$

Consequently

$$D_{un} = [\frac{y_{\min}(1 - a_1)}{b_1 c_1}, \frac{y_{\max}(1 - a_2)}{b_2 c_2}].$$

The final result is then as follows: If

$$\frac{y_{\min}}{b_1 c_1} - \frac{a_1 s_{01}}{b_1} \leq u_0 \leq \frac{y_{\max}}{b_2 c_2} - \frac{a_2 s_{02}}{b_2} \tag{5.5}$$

and for every $n > 0$

$$\frac{y_{\min}(1 - a_1)}{b_1 c_1} \leq u_n \leq \frac{y_{\max}(1 - a_2)}{b_2 c_2} \tag{5.6}$$

then the requirement concerning y_n will be satisfied. The conditions for the existence of the solution are the following:

$$\frac{y_{\min}}{b_1 c_1} - \frac{a_1 s_{01}}{b_1} \leq \frac{y_{\max}}{b_2 c_2} - \frac{a_2 s_{02}}{b_2}, \tag{5.7}$$

$$\frac{y_{\min}}{c_1} \leq \frac{y_{\max}}{c_2}, \tag{5.8}$$

$$\frac{y_{\min}(1 - a_1)}{b_1 c_1} \leq \frac{y_{\max}(1 - a_2)}{b_2 c_2}. \tag{5.9}$$

If $y_{\min} > 0$ and $a_2 < 1$ then these conditions are reduced to the inequality

$$\frac{y_{\max}}{y_{\min}} \geq \max(\alpha, \beta)$$

where

$$\alpha = \frac{b_2 c_2}{b_1 c_1} \cdot \frac{1 - a_1}{1 - a_2}, \qquad \beta = \frac{c_2}{c_1}.$$

Then D_{un} are not empty sets if the requirement concerning y_n is not too strong, i.e. the ratio $y_{\max} \cdot y_{\min}^{-1}$ is respectively high. One should note that the inequalities (5.7), (5.8), (5.9) form the sufficient condition for the existence of the sequence u_n satisfying the requirement.

5.2 Analysis and Decision Making for the Dynamical Plants with Uncertain Parameters

The analysis and decision problems for dynamical plants described by a relational knowledge representation with uncertain parameters may be formulated and solved in a similar way as for the static plants in Sects. 3.4 and 3.5. Let us consider the plant described by relations

$$\left.\begin{aligned} R_I(u_n, s_n, s_{n+1}; x) &\subseteq U \times S \times S, \\ R_{II}(s_n, y_n; w) &\subseteq S \times Y, \end{aligned}\right\} \tag{5.10}$$

where $x \in X$ and $w \in W$ are unknown vector parameters which are assumed to be values of uncertain variables $(\bar{x}, \bar{w})$ with the joint certainty distributions $h(x, w)$. We shall consider the analysis and decision problems in version II (see Sects 3.4 and 3.5) which has better practical interpretation. The considerations in version I are analogous.

Analysis problem: For the given relations (5.10), $h(x, w)$, D_{s0} and the sequences D_{un}, D_{yn} one should determine

$$v[D_{yn}(\bar{x}, \bar{w}) \tilde{\subseteq} D_{yn}] \overset{\Delta}{=} v_n$$

where $D_{yn}(\bar{x}, \bar{w})$ is the result of the analysis problem formulated in the previous section, i.e. the set of all possible outputs y_n for the fixed x and w.

In a similar way as for the static plant considered in Sect. 3.4 (see the formulas (3.33) and (3.34)) we obtain

$$v_n = v[(\bar{x}, \bar{w}) \tilde{\in} D(D_{yn}, D_{u,n-1})] = \max_{(x,w) \in D(D_{yn}, D_{u,n-1})} h(x, w) \tag{5.11}$$

where

$$D(D_{yn}, D_{u,n-1}) = \{(x, w) \in X \times W : D_{yn}(x, w) \subseteq D_{yn}\}.$$

In the case where $(\bar{x}, \bar{w})$ are considered as C-uncertain variables it is necessary to find v_n (5.11) and

$$v[(\bar{x}, \bar{v}) \tilde{\in} \bar{D}(D_{yn}, D_{u,n-1})] = \max_{(x,w) \in \bar{D}(D_{yn}, D_{u,n-1})} h(x, w)$$

where $\overline{D}(D_{yn}, D_{u,n-1}) = X \times W - D(D_{yn}, D_{u,n-1})$. Then

$$v_c[D_{yn}(\bar{x}, \bar{w}) \tilde{\subseteq} D_{yn}] = \frac{1}{2}\{v[(\bar{x}, \bar{w}) \tilde{\in} \overline{D}(D_{yn}, D_{u,n-1})]$$
$$+ 1 - v[(\bar{x}, \bar{w}) \tilde{\in} \overline{D}(D_{yn}, D_{u,n-1})]\}$$

(see (3.35) and (3.36)).

For the given value u_n, using (5.3) and (5.4) for the fixed (x, w) we obtain

$$D_{s,n+1}(u_n; x) = \{s_{n+1} \in S : \bigvee_{s_n \in D_{sn}(x)} (u_n, s_n, s_{n+1}) \in R_I(u_n, s_n, s_{n+1}; x)\},$$

$$D_{y,n+1}(u_n; x, w) = \{y_{n+1} \in Y : \bigvee_{s_{n+1} \in D_{s,n+1}(x)} [(s_{n+1}, y_{n+1}) \in R_{II}(s_{n+1}, y_{n+1}; w)]\}.$$

$$(5.12)$$

The formulation and solution of the analysis problem are the same as described above with u_n, $D_{yn}(u_{n-1}; x, w)$ and $D(D_{yn}, u_{n-1})$ instead of D_{un}, $D_{yn}(x, w)$ and $D(D_{yn}, D_{u,n-1})$, respectively.

Decision problem: For the given relations (5.10), $h(x, w)$, D_{s0} and the sequence D_{yn} $(n = 1, 2, ..., N)$ find the sequence of the optimal decisions

$$\hat{u}_n = \arg \max_{u_n \in U} v[D_{y,n+1}(u_n; x, w) \tilde{\subseteq} D_{y,n+1}]$$

for $n = 0, 1, ..., N - 1$, where $D_{y,n+1}(u_n; x, w)$ is the result of the analysis problem (5.12). Then

$$\hat{u}_n = \arg \max_{u_n \in U} \max_{(x,w) \in D(D_{yn}, u_{n-1})} h(x, w)$$

where

$$D(D_{yn}, u_{n-1}) = \{(x, w) \in X \times W : D_{yn}(u_{n-1}; x, w) \subseteq D_{yn}\}.$$

In the similar way as in Sect. 3.5 the determination of $\hat{u}_n$ may be replaced by the determination of $\hat{u}_{dn} = \hat{u}_n$ where

$$\hat{u}_{nd} = \arg \max_{u_n \in U} v[u_n \tilde{\in} D_{un}(\bar{x}, \bar{w})]$$

where $D_{un}(\bar{x}, \bar{w})$ is the result of the decision problem considered in the previous section for the fixed x and w. Then

$$\hat{u}_{nd} = \hat{u}_n = \arg\max_{u_n \in U} \max_{(x,w) \in D_d(u_n)} h(x, w)$$

where

$$D_d(u_n) = \{(x, w) : u_n \in D_{un}(x, w)\}.$$

Example 5.2: Let us assume that in the plant considered in Example 5.1 the parameters $c_1 \triangleq x_1$ and $c_2 \triangleq x_2$ are unknown and are the values of independent uncertain variables $\bar{x}_1$ and $\bar{x}_2$, respectively. The certainty distributions $h_{x1}(x_1)$ and $h_{x2}(x_2)$ have the triangular form with the parameters d_1, γ_1 for $\bar{x}_1$ (Fig. 5.2) and d_2, γ_2 for $\bar{x}_2$; $\gamma_1 < d_1$, $\gamma_2 < d_2$. Using the results (5.5) and (5.6) one may determine the optimal decisions $\hat{u}_n$, maximizing the certainty index $v[u_n \tilde{\in} D_{un}(\bar{x}_1, \bar{x}_2)] = v(u_n)$. From (5.6) we have

$$\begin{aligned}
v(u_n) &= v\{[\bar{x}_1 \tilde{\in} D_1(u_n)] \wedge [\bar{x}_2 \tilde{\in} D_2(u_n)]\} \\
&= \min\{v[\bar{x}_1 \tilde{\in} D_1(u_n)], v[\bar{x}_2 \tilde{\in} D_2(u_n)]\}.
\end{aligned} \tag{5.13}$$

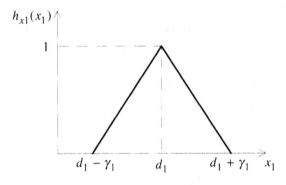

Fig. 5.2. Example of certainty distribution

Under assumption $a_{1,2} < 1$, the sets $D_1(u_n)$ and $D_2(u_n)$ for $n > 0$ are determined by the inequalities

$$x_1 \geq \frac{\alpha}{u_n}, \quad x_2 \leq \frac{\beta}{u_n}, \tag{5.14}$$

respectively, where

$$\alpha = \frac{y_{\min}(1 - a_1)}{b_1}, \quad \beta = \frac{y_{\max}(1 - a_2)}{b_2}.$$

The certainty indexes

$$v[\bar{x}_1 \,\tilde{\in}\, D_1(u_n)] = \max_{x_1 \in D_1(u_n)} h_{x1}(x_1) \overset{\Delta}{=} v_1(u_n)$$

and

$$v[\bar{x}_2 \,\tilde{\in}\, D_2(u_n)] = \max_{x_2 \in D_2(u_n)} h_{x2}(x_2) \overset{\Delta}{=} v_2(n)$$

may be obtained by using (5.14) and h_{x1}, h_{x2}:

$$v_1(u_n) = \begin{cases} 1 & \text{for} \quad u_n \geq \dfrac{\alpha}{d_1} \\[2ex] -\dfrac{\alpha}{u_n \gamma_1} + 1 + \dfrac{d_1}{\gamma_2} & \text{for} \quad \dfrac{\alpha}{d_1 + \gamma_1} \leq u_n \leq \dfrac{\alpha}{d_1} \\[2ex] 0 & \text{for} \quad u_n \leq \dfrac{\alpha}{d_1 + \gamma} \end{cases},$$

$$v_2(u_n) = \begin{cases} 1 & \text{for} \quad u_n \leq \dfrac{\beta}{d_2} \\[2ex] \dfrac{\beta}{\gamma_2 u_n} + 1 - \dfrac{d_2}{\gamma_2} & \text{for} \quad \dfrac{\beta}{d_2} \leq u_n \leq \dfrac{\beta}{d_2 - \gamma_2} \\[2ex] 0 & \text{for} \quad u_n \geq \dfrac{\beta}{d_2 - \gamma_2} \end{cases}.$$

Now we can consider three cases illustrated in Figs. 5.3, 5.4 and 5.5:

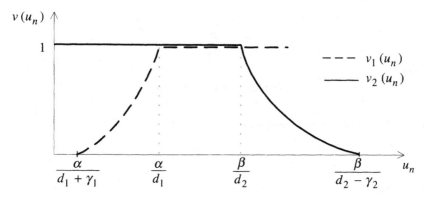

Fig. 5.3. Relationship between v and u – the first case

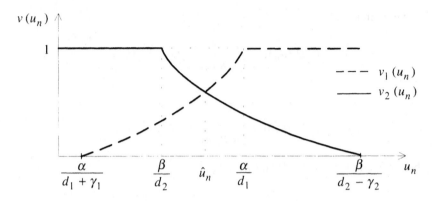

Fig. 5.4. Relationship between v and u – the second case

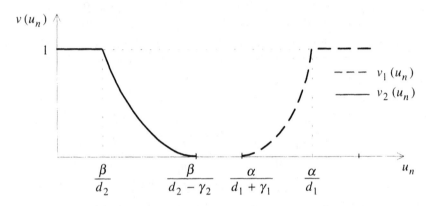

Fig. 5.5. Relationship between v and u – the third case

1.
$$\frac{\alpha}{d_1} \le \frac{\beta}{d_2}.$$

Then
$$\hat{u}_n = \arg\max_{u_n} \min\{v_1(u_n), v_2(u_n)\}$$

is any value satisfying the inequality

$$\frac{\alpha}{d_1} \le u_n \le \frac{\beta}{d_2}$$

and $v(\hat{u}_n) = 1$.

2.
$$\frac{\alpha}{d_1} > \frac{\beta}{d_2}, \quad \frac{\alpha}{d_1 + \gamma_1} < \frac{\beta}{d_2 - \gamma_2}.$$

Then

$$\hat{u}_n = \arg \max_{u_n} \min\{v_1(u_n), v_2(u_n)\} = \frac{\gamma_1 \beta + \gamma_2 \alpha}{\gamma_1 d_2 + \gamma_2 d_1}$$

and

$$v(\hat{u}_n) = \frac{\beta d_1 - \alpha d_2}{\beta \gamma_1 + \alpha \gamma_2} + 1.$$

3.

$$\frac{\alpha}{d_1 + \gamma} \geq \frac{\beta}{d_2 - \gamma_2}$$

Then for every u_n

$$v(u_n) = \min\{v_1(u_n), v_2(u_n)\} = 0$$

which means that the decision for which the requirement is satisfied with the certainty index greater than 0 does not exist.

The results for u_0 based on the inequality (5.5) have a similar form. It is important to note that the results are correct under the assumption

$$\frac{y_{min}}{d_1 - \gamma_1} \leq \frac{y_{max}}{d_2 + \gamma_2}$$

which means that the condition (5.8) is satisfied for every x_1 and x_2. Otherwise $v(u_n)$ may be smaller:

$$v(u_n) = \min\{v_1(u_n), v_2(u_n), v_3\}$$

where v_3 is the certainty index that the condition

$$\frac{y_{min}}{x_1} \leq \frac{y_{max}}{x_2}$$

is approximately satisfied, i.e.

$$v_3 = v(\frac{x_2}{x_1} \tilde{\in} D)$$

where $D \subset R^1 \times R^1$ is determined by the inequality

$$\frac{x_2}{x_1} \leq \frac{y_{max}}{y_{min}}.$$

5.3 Closed-Loop Control System. Uncertain Controller [17, 34]

The approach based on uncertain variables may be applied to closed-loop control systems containing continuous dynamic plant with unknown parameters which are assumed to be values of uncertain variables. The plant may be described by a classical model or by a relational knowledge representation. Now let us consider two control algorithms for the classical model of the plant, analogous to the algorithms Ψ and Ψ_d presented in Sect. 2.3: the control algorithm based on KP and the control algorithm based on KD which may be obtained from KP or may be given directly by an expert. The plant is described by the equations

$$\dot{s}(t) = f[s(t), u(t); x],$$
$$y(t) = \eta[s(t)]$$

where s is a state vector, or by the transfer function $K_P(p; x)$ in the linear case. The controller with the input y (or the control error ε) is described by the analogous model with a vector of parameters b which is to be determined. Consequently, the performance index

$$Q = \int_0^T \varphi(y, u)\, dt \overset{\Delta}{=} \Phi(b, x)$$

for the given T and φ is a function of b and x. In particular, for one-dimensional plant

$$Q = \int_0^\infty \varepsilon^2(t)\, dt = \Phi(b, x).$$

The closed-loop control system is then considered as a static plant with the input b, the output Q and the unknown parameter x, for which we can formulate and solve the decision problem described in Sect. 2.2. The control problem consisting in the determination of b in the known form of the control algorithm may be formulated as follows.

Control problem: For the given models of the plant and the controller find the value $\hat{b}$ minimizing $M(\overline{Q})$, i.e. the mean value of the performance index.

The procedure for solving the problem is then the following:

1. To determine the function $Q = \Phi(b, x)$.

2. To determine the certainty distribution $h_q(q; b)$ for $\overline{Q}$ using the function Φ and the distribution $h_x(x)$ in the same way as in the formula (2.1) for $\overline{y}$.

3. To determine the mean value $M(\overline{Q}; b)$.

4. To find $\hat{b}$ minimizing $M(\overline{Q};b)$.

In the second approach corresponding to the determination of Ψ_d for the static plant, it is necessary to find the value $b(x)$ minimizing $Q = \Phi(b, x)$ for the fixed x. The control algorithm with the uncertain parameter $b(x)$ may be considered as a knowledge of the control in our case, and the controller with this parameter may be called an *uncertain controller*. The deterministic control algorithm may be obtained in two ways, giving the different results. The first way consists in substituting $M(\overline{b})$ in the place of $b(x)$ in the uncertain control algorithm, where $M(\overline{b})$ should be determined using the function $b(x)$ and the certainty distribution $h_x(x)$. The second way consists in determination of the relationship between $u_d = M(\overline{u})$ and the input of the controller, using the form of the uncertain control algorithm and the certainty distribution $h_x(x)$. This may be very difficult for the dynamic controller.

The problem may be easier if the state of the plant $s(t)$ is put at the input of the controller. Then the uncertain controller has the form

$$u = \Psi(s, x)$$

which may be obtained as a result of nonparametric optimization, i.e. Ψ is the optimal control algorithm for the given model of the plant with the fixed x and for the given form of a performance index. Then

$$u_d = M(\overline{u}; s) \triangleq \Psi_d(s)$$

where $M(\overline{u}; s)$ is determined using the distribution

$$h_u(u; s) = v[\overline{x} \,\tilde{\in}\, D_x(u; s)] = \max_{x \in D_x(u;s)} h_x(x)$$

and

$$D_x(u; s) = \{x \in X : u = \Psi(s, x)\}.$$

5.4 Examples

Example 5.3: The data for the linear control system under consideration (Fig. 5.6) are the following:

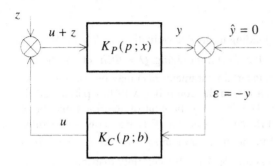

Fig. 5.6. Closed-loop control system

$$K_P(p;x) = \frac{x}{(pT_1 + 1)(pT_2 + 1)}, \quad K_C(p;b) = \frac{b}{p}$$

$z(t) = 0$ for $t < 0$, $z(t) = 1$ for $t \geq 0$, $h_x(x)$ has a triangular form presented in

Fig. 5.2 with $d_1 \overset{\Delta}{=} a$ and $\gamma_1 \overset{\Delta}{=} d$.

It is easy to determine

$$Q = \int_0^\infty \varepsilon^2(t)\, dt = \frac{x^2(T_1 + T_2)}{2xb(T_1 + T_2 - xbT_1T_2)} = \Phi(b, x). \qquad (5.15)$$

The minimization of Q with respect to b gives

$$b(x) = \frac{\alpha}{x}, \qquad \alpha = \frac{T_1 + T_2}{T_1T_2},$$

i.e. the uncertain controller is described by

$$K_C(p) = \frac{b(x)}{p} = \frac{\alpha}{xp}.$$

The certainty distribution $h_b(b)$ is as follows:

$$h_b(b) = \begin{cases} 0 & \text{for} \quad 0 < b \leq \dfrac{\alpha}{a + d} \\[2mm] \dfrac{ab - \alpha}{db} + 1 & \text{for} \quad \dfrac{\alpha}{a + d} \leq b \leq \dfrac{\alpha}{a} \\[2mm] \dfrac{-ab + \alpha}{db} + 1 & \text{for} \quad \dfrac{\alpha}{a} \leq b \leq \dfrac{\alpha}{a - d} \\[2mm] 0 & \text{for} \quad \dfrac{\alpha}{a - d} \leq b < \infty \end{cases}.$$

From the definition of a mean value we obtain

$$M(\bar{b}) = \frac{\alpha d(d^2 + 2a^2)}{2a^2 \ln \dfrac{a^2}{a^2 - d^2}} . \qquad (5.16)$$

Finally, the deterministic controller is described by

$$K_{C,d}(p) = \frac{M(\bar{b})}{p} .$$

To apply the first approach described in the previous section, it is necessary to find the certainty distribution for Q using formula (5.15) and the distribution $h_x(x)$, then to determine $M(\overline{Q}; b)$ and to find the value $\hat{b}$ minimizing $M(\overline{Q}; b)$. It may be shown that $\hat{b} \ne M(\bar{b})$ given by the formula (5.16).

Example 5.4: Let us consider the time-optimal control of the plant with $K_P(p;x) = xp^{-2}$ (Fig. 5.7), subject to constraint $|u(t)| \le M$.

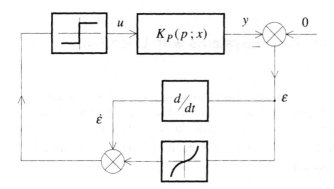

Fig. 5.7. Example of control system

It is well known that the optimal control algorithm is the following

$$u(t) = M \operatorname{sgn}(\varepsilon + |\dot{\varepsilon}|\dot{\varepsilon}(2xM)^{-1}) .$$

For the given $h_x(x)$ we can determine $h_u(u; \varepsilon, \dot{\varepsilon})$ which is reduced to three values $v_1 = v(\bar{u} \cong M)$, $v_2 = v(\bar{u} \cong -M)$, $v_3 = v(\bar{u} \cong 0)$. Then

$$u_d(t) = M(\bar{u}) = M(v_1 - v_2)(v_1 + v_2 + v_3)^{-1} .$$

It is easy to see that

$$v_1 = \max_{x \in D_{x1}} h_x(x), \qquad v_2 = \max_{x \in D_{x2}} h_x(x)$$

where

$$D_{x1} = \{x : x \operatorname{sgn} \varepsilon > -|\dot{\varepsilon}|\dot{\varepsilon}(2M|\varepsilon|)^{-1}\},$$

$$D_{x2} = \{x : x \operatorname{sgn} \varepsilon < -|\dot{\varepsilon}|\dot{\varepsilon}(2M|\varepsilon|)^{-1}\}$$

and $v_3 = h_x(-|\dot{\varepsilon}|\dot{\varepsilon}(2M\varepsilon)^{-1})$.

Assume that the certainty distribution of $\bar{x}$ is the same as in Example 5.3. For $\varepsilon > 0$, $\dot{\varepsilon} < 0$ and $x_g < a$ it is easy to obtain the following control algorithm

$$u_d = M(\bar{u}) = \begin{cases} M & \text{for} \quad d \leq a - x_g \\ M\dfrac{a - x_g}{3d - 2(a - x_g)} & \text{for} \quad d \geq a - x_g \end{cases}$$

where $x_g = -|\dot{\varepsilon}|\dot{\varepsilon}(2M\varepsilon)^{-1}$. For example for $M = 0.5$, $\dot{\varepsilon} = -3$, $\varepsilon = 1$, $a = 16$ and $d = 10$ we obtain $u_d = 0.2$.

5.5 Stability of Dynamical Systems with Uncertain Parameters

The uncertain variables and certainty distributions may be used in a qualitative analysis which consists in proving if the system with uncertain parameter $\bar{x}$ satisfies a determined property $P(x)$. In such a case, knowing the certainty distribution $h_x(x)$ we can calculate the certainty index that the property under consideration is approximately satisfied:

$$v(\bar{x} \tilde{\in} D_x) = \max_{x \in D_x} h_x(x)$$

where

$$D_x = \{x \in X : P(x)\}$$

(see (1.3)). Let us explain it for the stability of the discrete system

$$s_{n+1} = f(s_n, x) \tag{5.17}$$

where $s_n \in S = \mathbf{R}^k$ is a state vector and $x \in X$ is an unknown vector parameter which is assumed to be a value of an uncertain variable described by the known certainty distribution $h_x(x)$. The considerations are analogous to those for the system with random parameters presented in [21, 27]. Let us assume that the sys-

tem (5.17) has one equilibrium state equal to $\bar{0}$ (a vector with zero components).
For the linear time-invariant system

$$s_{n+1} = A(x)s_n$$

the necessary and sufficient condition of stability is as follows

$$\bigwedge_{i \in \overline{1,k}} |\lambda_i[A(x)]| < 1 \tag{5.18}$$

where λ_i denotes an eigenvalue of the matrix $A(x)$. The inequality (5.18) is a property $P(x)$ in this case and the certainty index v_s that the system is stable may be obtained in the following way

$$v_s = \max_{x \in D_x} h_x(x)$$

where

$$D_x = \{x \in X : \bigwedge_{i \in \overline{1,k}} |\lambda_i[A(x)]| < 1 \}.$$

Consider the nonlinear and/or time-varying system described by

$$s_{n+1} = A(s_n, c_n, x)s_n \tag{5.19}$$

where $c_n \in C$ is a vector of time-varying parameters and the uncertainty concerning c_n is formulated as follows

$$\bigwedge_{n \geq 0} c_n \in D_c \tag{5.20}$$

where D_c is a given set in C. The system (5.19) is globally asymptotically stable (GAS) iff s_n converges to $\bar{0}$ for any s_0. For the fixed x, the uncertain system (5.19), (5.20) is GAS iff the system (5.19) is GAS for every sequence c_n satisfying (5.20). Let $W(x)$ and $P(x)$ denote properties concerning x such that $W(x)$ is a sufficient condition and $P(x)$ is a necessary condition of the global asymptotic stability for the system (5.19), (5.20), i.e.

$$W(x) \rightarrow \text{the system (5.19), (5.20) is GAS,}$$

$$\text{the system (5.19), (5.20) is GAS} \rightarrow P(x).$$

Then the certainty index v_s that the system (5.19), (5.20) is GAS may be estimated by the inequality

$$v_w \leq v_s \leq v_p$$

where

$$v_w = \max_{x \in D_{xw}} h_x(x), \qquad v_p = \max_{x \in D_{xp}} h_x(x),$$

$$D_{xw} = \{x \in X : W(x)\}, \qquad D_{xp} = \{x \in X : P(x)\},$$

v_w is the certainty index that the sufficient condition is approximately satisfied and v_p is the certainty index that the necessary condition is approximately satisfied. In general, $D_{xw} \subseteq D_{xp}$ and $D_{xp} - D_{xw}$ may be called "a grey zone" which is a result of an additional uncertainty caused by the fact that $W(x) \neq P(x)$. In particular, if it is possible to determine a sufficient and necessary condition $W(x) = P(x)$ then $v_w = v_p$ and the value v_s may be determined exactly. The condition $P(x)$ may be determined as a negation of a sufficient condition that the system is not GAS, i.e. such a property $P_{neg}(x)$ that

$$P_{neg}(x) \rightarrow \text{ there exists } c_n \text{ satisfying (5.20) such that (5.19) is not GAS.}$$

For the nonlinear and time-varying system we may use the stability conditions in the form of the following theorems presented in [4, 5, 6, 21]:

Theorem 5.1: If there exists a norm $\| \cdot \|$ such that

$$\bigwedge_{c \in D_c} \bigwedge_{s \in S} \| A(s, c, x) \| < 1$$

then the system (5.19), (5.20) is GAS.   □

The final form of the set

$$D_{xw} = \{x \in X : \bigwedge_{c \in C} \bigwedge_{s \in S} \| A(s, c, x) \| < 1\}$$

depends on the form of the norm. In particular the norm $\| A \|$ may have the form

$$\| A \|_2 = \sqrt{\lambda_{max}(A^T A)} \tag{5.21}$$

where λ_{max} is the maximum eigenvalue of the matrix $A^T A$,

$$\| A \|_1 = \max_{1 \le i \le k} \sum_{j=1}^{k} | a_{ij} |, \qquad \| A \|_\infty = \max_{1 \le j \le k} \sum_{i=1}^{k} | a_{ij} |. \qquad (5.22)$$

Theorem 5.2:   Consider a linear, time-varying system

$$s_{n+1} = A(c_n, x)s_n. \qquad (5.23)$$

If the system (5.23), (5.20) is GAS then

$$\bigwedge_{c \in D_c} \max_i | \lambda_i [A(c, x)] | < 1 \qquad (5.24)$$

where $\lambda_i(A)$ are the eigenvalues of the matrix A ($i = 1, 2, ..., k$).   □

Theorem 5.3:   The system (5.19), (5.20) where

$$D_c = \{c \in C : \bigwedge_{s \in S} [\underline{A}(x) \le A(s, c, x) \le \overline{A}(x)]\} \qquad (5.25)$$

is GAS if all entries of the matrices $\underline{A}(x)$ and $\overline{A}(x)$ are nonnegative and

$$\| \overline{A}(x) \| < 1. \qquad (5.26)$$

□

The inequality in (5.25) denotes the inequalities for the entries:

$$\underline{a}_{ij}(x) \le a_{ij}(s, c, x) \le \overline{a}_{ij}(x).$$

Theorem 5.4:   Assume that all entries of the matrix $\underline{A}(x)$ are nonnegative. If the system (5.19), (5.25) is GAS then

$$\bigvee_j \sum_{i=1}^{k} \underline{a}_{ij}(x) < 1 \qquad (5.27)$$

and

$$\bigvee_i \sum_{j=1}^{k} \underline{a}_{ij}(x) < 1. \qquad (5.28)$$

□

Choosing different sufficient and necessary conditions we may obtain the different estimations of the certainty index v_s. For example, if we choose the condition (5.26) with the norm $\| \cdot \|_\infty$ in (5.22) and the condition (5.27) then

$$D_{xw} = \{ x \in X : \bigwedge_j \sum_{i=1}^{k} \bar{a}_{ij}(x) < 1 \}, \tag{5.29}$$

$$D_{xp} = X - D_{x,\text{neg}}$$

where

$$D_{x,\text{neg}} = \{ x \in X : \bigwedge_j \sum_{i=1}^{k} \underline{a}_{ij}(x) \geq 1 \}. \tag{5.30}$$

Example 5.5: Consider an uncertain system (5.19) where $k = 2$ and

$$A(s_n, c_n, x) = \begin{bmatrix} a_{11}(s_n, c_n) + x & a_{12}(s_n, c_n) \\ a_{21}(s_n, c_n) & a_{22}(s_n, c_n) + x \end{bmatrix}$$

with the uncertainty (5.25), i.e. nonlinearities and the sequence c_n are such that

$$\bigwedge_{c \in D_c} \bigwedge_{s \in D_s} \underline{a}_{ij} \leq a_{ij}(s, c) \leq \bar{a}_{ij}, \quad i = 1, 2; \ j = 1, 2.$$

Assume that $x \geq 0$ and $\underline{a}_{ij} \geq 0$. Applying the condition (5.26) with the norm $\| \cdot \|_{\infty}$ in (5.22) yields

$$\bar{a}_{11} + x + \bar{a}_{21} < 1, \quad \bar{a}_{12} + \bar{a}_{22} + x < 1$$

and D_{xw} in (5.29) is defined by

$$x < 1 - \max(\bar{a}_{11} + \bar{a}_{21}, \bar{a}_{12} + \bar{a}_{22}).$$

Applying the negation of the condition (5.27) yields

$$\underline{a}_{11} + x + \underline{a}_{21} \geq 1, \quad \underline{a}_{12} + \underline{a}_{22} + x \geq 1.$$

Then $D_{x,\text{neg}}$ in (5.30) is determined by

$$x \geq 1 - \min(\underline{a}_{11} + \underline{a}_{21}, \underline{a}_{12} + \underline{a}_{22})$$

and the necessary condition (5.27) defining the set $D_{xp} = X - D_{x,\text{neg}}$ is as follows

$$x < 1 - \min(\underline{a}_{11} + \underline{a}_{21}, \underline{a}_{12} + \underline{a}_{22}).$$

For the given certainty distribution $h_x(x)$ we can determine

$$v_w = \max_{0 \le x \le x_w} h_x(x), \qquad v_p = \max_{0 \le x \le x_p} h_x(x) \qquad (5.31)$$

where

$$x_w = 1 - \max(\bar{a}_{11} + \bar{a}_{21}, \bar{a}_{12} + \bar{a}_{22}),$$
$$x_p = 1 - \min(\underline{a}_{11} + \underline{a}_{21}, \underline{a}_{12} + \underline{a}_{22}).$$

Assume that $h_x(x)$ has triangular form presented in Fig. 5.8. The results obtained from (5.31) for the different cases are as follows:

1. For $x_w \ge d + \gamma$

$$v_w = v_p = 1.$$

2. For $d \le x_w \le d + \gamma$

$$v_w = -\frac{x_w}{\gamma} + 1 + \frac{d}{\gamma} \triangleq v_1,$$

$$v_p = \begin{cases} 1 & \text{for } x_p \ge d + \gamma \\ -\dfrac{x_p}{\gamma} + 1 + \dfrac{d}{\gamma} & \text{otherwise} \end{cases}.$$

3. For $d - \gamma \le x_w \le d$

$$v_w = \frac{x_w}{\gamma} + 1 - \frac{d}{\gamma},$$

$$v_p = \begin{cases} 1 & \text{for } x_p \ge d + \gamma \\ -\dfrac{x_p}{\gamma} + 1 + \dfrac{d}{\gamma} & \text{for } d \le x_p \le d + \gamma \\ \dfrac{x_p}{\gamma} + 1 - \dfrac{d}{\gamma} & \text{otherwise} \end{cases}.$$

4. For $x_w \le d - \gamma$

$$v_w = 0,$$

$$v_p = \begin{cases} 1 & \text{for } x_p \geq d + \gamma \\ -\dfrac{x_p}{\gamma} + 1 + \dfrac{d}{\gamma} & \text{for } d \leq x_p \leq d + \gamma \\ \dfrac{x_p}{\gamma} + 1 - \dfrac{d}{\gamma} & \text{for } d - \gamma \leq x_p \leq d \\ 0 & \text{otherwise} \end{cases}$$

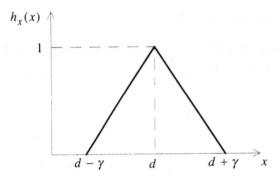

Fig. 5.8. Example of certainty distribution

For example, if $d - \gamma \leq x_w \leq d$ and $x_p \leq d$ then the certainty index v_s that the system is globally asymptotically stable satisfies the following inequality

$$\frac{x_w}{\gamma} + 1 - \frac{d}{\gamma} \leq v_s \leq \frac{x_p}{\gamma} + 1 - \frac{d}{\gamma} \, .$$

6 Comparison, Analogies and Generalisation

6.1 Comparison with Random Variables and Fuzzy Numbers

The formal part of the definitions of a random variable, a fuzzy number and an uncertain variable is the same: $< X, \mu(x) >$, that is a set X and a function $\mu : X \rightarrow R^1$ where $0 \leq \mu(x)$ for every $x \in X$. For the fuzzy number, the uncertain variable and for the random variable in the discrete case, $\mu(x) \leq 1$. For the random variable the property of additivity is required, which in the discrete case $X = \{x_1, x_2, ..., x_m\}$ is reduced to the equality $\mu(x_1) + \mu(x_2) + ... + \mu(x_m) = 1$. Without any additional description, one can say that each variable is defined by a fuzzy set $< X, \mu(x) >$. In fact, each definition contains an additional description of semantics which discriminates the respective variables. To compare the uncertain variables with probabilistic and fuzzy approaches, take into account the definitions for $X \subseteq R^1$, using Ω, ω and $g(\omega) = \bar{x}(\omega)$ introduced in Sect. 1.1. *The random variable* $\tilde{x}$ is defined by X and probability distribution $\mu(x) = F(x)$ (or probability density $f(x) = F'(x)$ if this exists) where $F(x)$ is the probability that $\tilde{x} \leq x$. In discrete case $\mu(x_i) = p(x_i) = P(\tilde{x} = x_i)$ (probability that $\tilde{x} = x_i$). For example, if Ω is a set of 100 persons and 20 of them have the age $\bar{x}(\omega) = 30$, then the probability that a person chosen randomly from Ω has $\bar{x} = 30$ is equal to 0.2. In general, the function $p(x)$ (or $f(x)$ in a continuous case) is an *objective* characteristic of Ω as a whole and $h_\omega(x)$ is a subjective characteristic given by an expert and describing his or her individual opinion of the fixed particular ω.

To compare uncertain variables with fuzzy numbers, let us recall three basic definitions of the fuzzy number in a wide sense of the word, that is the definitions of the fuzzy set based on the number set $X = R^1$.

1. The fuzzy number $\hat{x}(d)$ for the given fixed value $d \in X$ is defined by X and the membership function $\mu(x, d)$ which may be considered as a logic value (*degree of truth*) of the soft property "if $\hat{x} = x$ then $\hat{x} \cong d$ ".

2. The linguistic fuzzy variable $\hat{x}$ is defined by X and a set of membership functions $\mu_i(x)$ corresponding to different descriptions of the size of $\hat{x}$ (small, medium, large, etc.). For example, $\mu_1(x)$ is a logic value of the soft property "if $\hat{x} = x$ then $\hat{x}$ is small".

3. The fuzzy number $\hat{x}(\omega)$ (where $\omega \in \Omega$ was introduced at the beginning of Sect. 1.1) is defined by X and the membership function $\mu_\omega(x)$ which is a logic value (*degree of possibility*) of the soft property "it is possible that the value x is assigned to ω ".

In the first two definitions the membership function does not depend on ω; in the third case there is a family of membership functions (a family of fuzzy sets) for $\omega \in \Omega$. The difference between $\hat{x}(d)$ or the linguistic fuzzy variable $\hat{x}$ and the uncertain variable $\overline{x}(\omega)$ is quite evident. The variables $\hat{x}(\omega)$ and $\overline{x}(\omega)$ are formally defined in the same way by the fuzzy sets $< X, \mu_\omega(x) >$ and $< X, h_\omega(x) >$ respectively, but the interpretations of $\mu_\omega(x)$ and $h_\omega(x)$ are different. In the case of the uncertain variable there exists a function $\overline{x} = g(\omega)$, the value $\overline{x}$ is determined for the fixed ω but is unknown to an expert who formulates the degree of certainty that $\overline{x}(\omega) \cong x$ for the different values $x \in X$. In the case of $\hat{x}(\omega)$ the function g may not exist. Instead we have a property of the type "it is possible that $P(\omega, x)$ " (or, briefly, "it is possible that the value x is assigned to ω ") where $P(\omega, x)$ is such a property concerning ω and x for which it makes sense to use the words "it is possible". Then $\mu_\omega(x)$ for fixed ω means the degree of possibility for the different values $x \in X$ given by an expert. The example with persons and age is not adequate for this interpretation. In the popular example of the possibilistic approach $P(\omega, x) =$ "John (ω) ate x eggs at his breakfast".

From the point of view presented above, $\overline{x}(\omega)$ may be considered as a special case of $\hat{x}(\omega)$ (when the relation $P(\omega, x)$ is reduced to the function g), with a specific interpretation of $\mu_\omega(x) = h_\omega(x)$. A further difference is connected with the definitions of $w(\overline{x} \tilde{\in} D_x)$, $w(\overline{x} \tilde{\notin} D_x)$, $w(\overline{x} \tilde{\in} D_1 \vee \overline{x} \tilde{\in} D_2)$ and $w(\overline{x} \tilde{\in} D_1 \wedge \overline{x} \tilde{\in} D_2)$. The function $w(\overline{x} \tilde{\in} D_x) \overset{\Delta}{=} m(D_x)$ may be considered as a *measure* defined for the family of sets $D_x \subseteq X$. Two measures have been defined in the definitions of the uncertain variables: $v(\overline{x} \tilde{\in} D_x) \overset{\Delta}{=} \overline{m}(D_x)$

and $v_c(\bar{x} \tilde{\in} D_x) \stackrel{\Delta}{=} m_c(D_x)$. Let us recall the following special cases of fuzzy measures (see for example [41]) and their properties for every D_1, D_2.

1. If $m(D_x)$ is a *belief measure*, then
$$m(D_1 \cup D_2) \geq m(D_1) + m(D_2) - m(D_1 \cap D_2).$$

2. If $m(D_x)$ is a *plausibility measure*, then
$$m(D_1 \cap D_2) \leq m(D_1) + m(D_2) - m(D_1 \cup D_2).$$

3. A *necessity measure* is a belief measure for which
$$m(D_1 \cap D_2) = \min\{m(D_1), m(D_2)\}.$$

4. A *possibility measure* is a plausibility measure for which
$$m(D_1 \cup D_2) = \max\{m(D_1), m(D_2)\}.$$

Taking into account the properties of $\bar{m}$ and m_c presented in Definitions 1.5 and 1.6 and in Theorems 1.1, 1.2 and 1.3, 1.4, it is easy to see that $\bar{m}$ is a possibility measure, that $m_n \stackrel{\Delta}{=} 1 - v(\bar{x} \tilde{\in} \bar{D}_x)$ is a necessity measure and that m_c is neither a belief nor a plausibility measure. To prove this for the plausibility measure, it is enough to take Example 1.3 as a counter-example:

$$m_c(D_1 \cap D_2) = 0.3 > m_c(D_1) + m_c(D_2) - m_c(D_1 \cup D_2) = 0.4 + 0.6 - 0.9.$$

For the belief measure, it follows from (1.66) when D_1 and D_2 correspond to the upper case, and from the inequality $\bar{m}(D_1 \cup D_2) = \max\{\bar{m}(D_1), \bar{m}(D_2)\} < \bar{m}(D_1) + \bar{m}(D_2)$ for $D_1 \cap D_2 = \varnothing$.

The interpretation of the membership function $\mu(x)$ as a logic value w of a given soft property $P(x)$, that is $\mu(x) = w[P(x)]$, is especially important and necessary if we consider two fuzzy numbers (x, y) and a relation $R(x, y)$ or a function $y = f(x)$. Consequently, it is necessary if we formulate analysis and decision problems. The formal relationships (see for example [39, 40])

$$\mu_y(y) = \max_x [\mu_x(x) : f(x) = y]$$

for the function and

$$\mu_y(y) = \max_x [\mu_x(x) : (x, y) \in R]$$

for the relation do not determine evidently $P_y(y)$ for the given $P_x(x)$. If $\mu_x(x) = w[P_x(x)]$ where $P_x(x) = $ "if $\hat{x} = x$ then $\hat{x} \cong d$", then we can accept that $\mu_y(y) = w[P_y(y)]$ where $P_y(y) = $ "if $\hat{y} = y$ then $\hat{y} \cong f(\hat{x})$" in the case of the function, but in the case of the relation $P_y(y)$ is not determined. If

$P_x(x) =$ "if $\hat{x} = x$ then $\hat{x}$ is small" , then $P_y(y)$ may not be evident even in the case of the function, for example $y = \sin x$. For the uncertain variable $\mu_x(x) = h_x(x) = v(\bar{x} \cong x)$ with the definitions (1.53) – (1.56), the property $P_y(y)$ such that $\mu_y(y) = v[P_y(y)]$ is determined precisely: in the case of the function, $\mu_y(y) = h_y(y) = v(\bar{y} \cong y)$ and, in the case of the relation, $\mu_y(y)$ is the certainty index of the property $P_y(y) =$ " there exist $\bar{x}$ such that $(\bar{x}, \bar{y}) \tilde{\in} R(x, y)$ ".

Consequently, using uncertain variables it is possible not only to formulate the analysis and decision problems in the form considered in Chaps. 2 and 3 but also to define precisely the meaning of these formulations and solutions. This corresponds to the two parts of the definition of the uncertain logic mentioned in Sect. 1.1 after Theorem 1.2: a formal description and its interpretation. The remark concerning ω in this definition is also very important because it makes it possible to interpret precisely the source of the information about the unknown parameter $\bar{x}$ and the term "certainty index".

In the theory of fuzzy sets and systems there exist other formulations of analysis and decision problems (see for example [39]), different from those presented in this paper. The decision problem with a fuzzy goal is usually based on the given $\mu_y(y)$ as the logic value of the property " $\hat{y}$ is satisfactory" or related properties.

The statements of analysis and decision problems in Chap. 3 for the system with the *known relation R and unknown parameter x* considered as an uncertain variable are similar to analogous approaches for the probabilistic model and together with the deterministic case form a unified set of problems. For $y = \Phi(u, x)$ and given y the decision problem is as follows.

1. If x is known (the deterministic case), find u such that $\Phi(u, x) = y$.

2. If x is a value of random variable $\tilde{x}$ with given certainty distribution, find u, maximizing the probability that $\tilde{y} = y$ (for the discrete variable), or find u such that $E(\tilde{y}, u) = y$ where E denotes the expected value of $\tilde{y}$.

3. If x is a value of uncertain variable $\bar{x}$ with given certainty distribution, find u, maximizing the certainty index of the property $\bar{y} \cong y$, or find u such that $M_y(u) = y$ where M denotes the mean value of $\bar{y}$.

The definition of the uncertain variable has been used to introduce an C-uncertain variable, especially recommended for analysis and decision problems with unknown parameters, because of its advantages mentioned in Sect. 1.3. Not only the interpretation but also a formal description of the C-uncertain variable differ in an obvious way from the known definitions of fuzzy numbers (see Definition 1.6 and the remark concerning the measure m_c in this section).

To indicate the analogies with the probabilistic approach and the approach based on the fuzzy description, in the next two sections we shall consider the non-parametric decision problems analogous to those presented in Sect. 2.4.

6.2 Application of Random Variables

Let us consider the static plant with the input $u \in U$, the output $y \in Y$ and the vector of external disturbances $z \in Z$ and let us assume that (u, y, z) are values of random variables $(\tilde{u}, \tilde{y}, \tilde{z})$. The knowledge of the plant given by an expert contains a conditional probability density $f_y(y \mid u, z)$ and the probability density $f_z(z)$ for $\tilde{z}$, i.e.

$$KP = < f_y(y \mid u, z), f_z(z) > .$$

Then it is possible to determine a *random decision algorithm* in the form of a conditional probability density $f_u(u \mid z)$, for the given desirable probability density $f_y(y)$ required by a user.

Decision problem: For the given $f_y(y \mid u, z)$, $f_z(z)$ and $f_y(y)$ one should determine $f_u(u \mid z)$.

The relationship between the probability densities $f_y(y)$ and $f_{uz}(u, z)$ is as follows

$$f_y(y) = \int_U \int_Z f_{uz}(u, z) f_y(y \mid u, z) du dz$$

where

$$f_{uz}(u, z) = f_z(z) f_u(u \mid z) \tag{6.1}$$

is the joint probability density for $(\tilde{u}, \tilde{z})$. Then

$$f_y(y) = \int_U \int_Z f_z(z) f_u(u \mid z) f_y(y \mid u, z) du dz . \tag{6.2}$$

Any probability density $f_u(u \mid z)$ satisfying the equation (6.2) is a solution of our decision problem. It is easy to note that the solution of the equation (6.2) may be not unique. Having $f_u(u \mid z)$ one can obtain the deterministic decision algorithm $\Psi(z)$ as a result of the determinization of the uncertain decision algorithm described by $f_u(u \mid z)$. Two versions corresponding to the formulation in (2.13) and (2.15) are the following:

I.

$$u_a = \arg \max_{u \in U} f_u(u \mid z) \triangleq \Psi_a(z).$$

II.

$$u_b = E(\tilde{u} \mid z) = \int_U u f_u(u \mid z) du \triangleq \Psi_b(z)$$

where E denotes the conditional expected value. The decision algorithms $\Psi_a(z)$ or $\Psi_b(z)$ are based on the knowledge of the decision making $KD = < f_u(u \mid z) >$ (or the random decision algorithm) which is determined from the knowledge of the plant KP for the given $f_y(y)$ (Fig. 6.1). The relationships (6.1) and (6.2) are analogous to (2.16) and (2.17) for the description using uncertain variables.

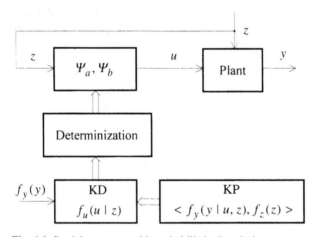

Fig. 6.1. Decision system with probabilistic description

6.3 Application of Fuzzy Numbers

Let us consider two fuzzy numbers defined by set of values $X \subseteq R^1$, $Y \subseteq R^1$ and membership functions $\mu_x(x)$, $\mu_y(y)$, respectively. The membership function $\mu_x(x)$ is the logic value of the soft property $\varphi_x(x) = $ "if $\hat{x} = x$ then $\hat{x}$ is d_1" or shortly "$\hat{x}$ is d_1", and $\mu_y(y)$ is the logic value of the soft property $\varphi_y(y) = $ "$\hat{y}$ is d_2", i.e.

$$w[\varphi_x(x)] = \mu_x(x), \quad w[\varphi_y(y)] = \mu_y(y)$$

where d_1 and d_2 denote the size of the number, e.g. $\varphi_x(x) = $ "$\hat{x}$ is small", $\varphi_y(y) = $ "$\hat{y}$ is large". Using the properties φ_x and φ_y we can introduce the property $\varphi_x \rightarrow \varphi_y$ (e.g. "if $\hat{x}$ is small then $\hat{y}$ is large") with the respective membership function

$$w[\varphi_x \rightarrow \varphi_y] \overset{\Delta}{=} \mu_y(y \mid x)$$

and the properties

$$\varphi_x \vee \varphi_y \quad \text{and} \quad \varphi_x \wedge \varphi_y = \varphi_x \wedge [\varphi_x \rightarrow \varphi_y]$$

for which the membership functions are defined as follows

$$w[\varphi_x \vee \varphi_y] = \max\{\mu_x(x), \mu_y(y)\},$$

$$w[\varphi_x \wedge \varphi_y] = \min\{\mu_x(x), \mu_y(y \mid x)\} \overset{\Delta}{=} \mu_{xy}(x, y). \tag{6.3}$$

If we assume that

$$\varphi_x \wedge [\varphi_x \rightarrow \varphi_y] = \varphi_y \wedge [\varphi_y \rightarrow \varphi_x]$$

then

$$\mu_{xy}(x, y) = \min\{\mu_x(x), \mu_y(y \mid x)\} = \min\{\mu_y(y), \mu_x(x \mid y)\}. \tag{6.4}$$

The properties φ_x, φ_y and the corresponding fuzzy numbers $\hat{x}$, $\hat{y}$ are called *independent* if

$$w[\varphi_x \wedge \varphi_y] = \mu_{xy}(x, y) = \min\{\mu_x(x), \mu_y(y)\}.$$

Using (6.4) it is easy to show that

$$\mu_x(x) = \max_{y \in Y} \mu_{xy}(x, y), \tag{6.5}$$

$$\mu_y(y) = \max_{x \in X} \mu_{xy}(x, y). \tag{6.6}$$

The equations (6.4) and (6.5) describe the relationships between μ_x, μ_y, μ_{xy}, $\mu_x(x \mid y)$ analogous to the relationships (1.76), (1.74), (1.75) for uncertain variables, in general defined in multidimensional sets X and Y. For the given $\mu_{xy}(x, y)$ the function $\mu_y(y \mid x)$ is determined by the equation (6.3) in which

$$\mu_x(x) = \max_{y \in Y} \mu_{xy}(x, y).$$

Theorem 6.1: The set of functions $\mu_y(y \mid x)$ satisfying the equation (6.3) is determined as follows:

$$\mu_y(y \mid x) \begin{cases} = \mu_{xy}(x, y) & \text{for} \quad (x, y) \notin D(x, y) \\ \geq \mu_{xy}(x, y) & \text{for} \quad (x, y) \in D(x, y) \end{cases}$$

where

$$D(x, y) = \{(x, y) \in X \times Y : \ \mu_x(x) = \mu_{xy}(x, y)\}.$$

Proof: From (6.3) it follows that

$$\bigwedge_{x \in X} \bigwedge_{y \in Y} [\mu_x(x) \geq \mu_{xy}(x, y)].$$

If $\mu_x(x) > \mu_{xy}(x, y)$ then, according to (6.3), $\mu_{xy}(x, y) = \mu_y(y \mid x)$. If $\mu_x(x) = \mu_{xy}(x, y)$, i.e. $(x, y) \in D(x, y)$ then $\mu_y(y \mid x) \geq \mu_{xy}(x, y)$.   □

In particular, as one of the solutions of the equation (6.3), i.e. one of the possible definition of the membership function for an implication we may accept

$$\mu_y(y \mid x) = \mu_{xy}(x, y).\tag{6.7}$$

If $\mu_{xy}(x, y) = \min\{\mu_x(x), \mu_y(y)\}$ then according to (6.7)

$$\mu_y(y \mid x) = \min\{\mu_x(x), \mu_y(y)\}$$

and according to (6.3)

$$\mu_y(y \mid x) = \mu_y(y).$$

The description concerning the pair of fuzzy numbers may be directly applied to one-dimensional static plant with one input $u \in U$, one disturbance $z \in Z$ and one output $y \in Y$ $(U, Z, Y \subseteq R^1)$. The nonparametric description of uncertainty using fuzzy numbers may be formulated by introducing three soft properties $\varphi_u(u)$, $\varphi_z(z)$ and $\varphi_y(y)$. This description is given by an expert and contains the membership function

$$w[\varphi_u \wedge \varphi_z \rightarrow \varphi_y] = \mu_y(y \mid u, z)$$

and the membership function $w[\varphi_z(z)] = \mu_z(z)$, i.e. the knowledge of the plant

$$KP = <\mu_y(y \mid u, z), \mu_z(z) >.$$

For examples, the expert says that "if $\hat{u}$ is large and $\hat{z}$ is medium then $\hat{y}$ is small" and gives the membership function $\mu_y(y \mid u, z)$ for this property and the membership function $\mu_z(z)$ for the property "$\hat{z}$ is medium". In this case the decision problem may consists in determination of such a membership function $\mu_u(u \mid z)$ for which the membership function for the output property $w[\varphi_y(y)] = \mu_y(y)$ will have a desirable form. In some sense, this is a problem analogous to that in the previous section for random variables and to that in Sect. 2.4 for uncertain variables. The essential difference consists in the fact that the requirements in the form of $h_y(y)$ or $f_y(y)$ have been concerned directly with a *value of the input* and now the requirement $\mu_y(y)$ concerns the *output property* $\varphi_y(y)$.

Decision problem: For the given $\mu_y(y \mid u, z)$, $\mu_z(z)$ and $\mu_y(y)$ one should determine $\mu_u(u \mid z)$.

Let us introduce $\mu_{u,z,y}(u, z, y) = w[\varphi_u \wedge \varphi_z \wedge \varphi_y]$. According to the general relationships (6.5) and (6.4)

$$\mu_y(y) = \max_{u \in U,\, z \in Z} \mu_{u,z,y}(u, z, y) = \max_{u,z} \min\{\mu_{uz}(u, z), \mu_y(y \mid u, z)\} \quad (6.8)$$

or

$$\mu_y(y) = \max_{u \in U,\, z \in Z} \min\{\mu_z(z), \mu_u(u \mid z), \mu_y(y \mid u, z)\}. \quad (6.9)$$

As a solution we may accept any function $\mu_u(u \mid z)$ satisfying the equation (6.9). The solution may be obtained in two steps. In the first step we determine the set of functions $\mu_{uz}(u, z)$ satisfying the equation (6.8) and in the second step we determine $\mu_u(u \mid z)$ from the equation

$$\mu_{uz}(u, z) = \min\{\mu_z(z), \mu_u(u \mid z)\}. \quad (6.10)$$

If the definition (6.7) is accepted then it is sufficient to determine $\mu_{uz}(u, z) = \mu_u(u \mid z)$ in the first step. It is easy to see that if the functions $\mu_y(y)$ and $\mu_y(y \mid u, z)$ have one local maximum equal to 1 then the point (u, z) maximizing the right hand side of the equation (6.8) satisfies the equation

$$\mu_{uz}(u, z) = \mu_y(y \mid u, z).$$

Hence, for this point we have

$$\mu_y(y) = \mu_y(y \mid u, z).$$ (6.11)

Consequently, the procedure for the determination of μ_{uz} analogous to that for the determination of h_{uz} presented in Sect. 2.4, is as follows:

1. To solve the equation (6.11) with respect to y and to obtain $y^*(u, z)$.

2. To put $y^*(u, z)$ into $\mu_y(y)$ in the place of y and to obtain

$$\mu_{uz}(u, z) = \mu_y[y^*(u, z)].$$

3. To assume $\mu_u(u \mid z) = \mu_{uz}(u, z)$ as one of the solutions of the equation (6.10).

The function $\mu_u(u \mid z)$ may be considered as the knowledge of the decision making $KD = < \mu_u(u \mid z) >$ or the *fuzzy decision algorithm* (fuzzy controller in open-loop control system). According to another version, the knowledge of the decision making is the function $\mu_{uz}(u, z)$ in (6.10), i.e. $KD = < \mu_u(u \mid z), \mu_z(z) >$. The determinization, i.e. the determination of the mean value, gives the deterministic decision algorithm

$$u_d = M(\hat{u}) = \Psi(z)$$

where the definition of the mean value $M(\hat{u})$ for a fuzzy number is the same as for an uncertain variable (see Sect. 1.4) with the membership function μ in the place of the certainty distribution h. Using $\mu_u(u \mid z)$ or $\mu_{uz}(u, z)$ with the fixed z in the determination of $M(\hat{u})$ one obtains two versions of $\Psi(z)$. The both versions are the same if we assume that $\mu_u(u \mid z) = \mu_{uz}(u, z)$. Let us note that in the analogous problems for uncertain variables (Sect. 2.4) and for random variables (Sect. 6.2) it is not possible to introduce two versions of KD considered here for fuzzy numbers. It is caused by the fact that $\mu_{uz}(u, z)$ and $\mu_u(u \mid z)$ does not concern directly the *values* of the variables (as probability distributions or certainty distributions) but are concerned with the properties φ_u, φ_z and

$$\mu_{uz}(u, z) = w[\varphi_u \wedge \varphi_z], \ \mu_u(u \mid z) = w[\varphi_z \rightarrow \varphi_u] = w[\varphi_u \mid \varphi_z].$$

The deterministic decision algorithm is based on the knowledge of the decision making KD which is determined from the knowledge of the plant KP for the given $\mu_u(u)$ (Fig. 6.2).

The considerations may be extended to multidimensional case with vectors u, y, z. To formulate the knowledge of the plant one introduces soft properties of the following form:

$$\varphi_{ui}(j) = \text{''} u^{(i)} \text{ is } d_j\text{''}, \qquad \varphi_{zi}(j) = \text{''} z^{(i)} \text{ is } d_j\text{''}, \qquad \varphi_{yi}(j) = \text{''} y^{(i)} \text{ is } d_j\text{''}$$

where $u^{(i)}$, $z^{(i)}$, $y^{(i)}$ denote i-th components of u, z, y, respectively and $d_j \in \{d_1, d_2, ..., d_m\}$ denotes the size of the number (e.g. small, medium, large, etc.). Each property is described by a membership function. Consequently, in the

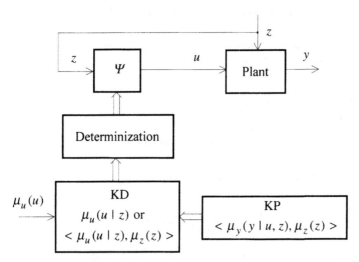

Fig. 6.2. Decision system with fuzzy description

place of one implication $\varphi_u \wedge \varphi_z \rightarrow \varphi_y$ now we have a set of implications for $\varphi_{uk}(l) \wedge \varphi_{zm}(p) \rightarrow \varphi_{yn}(s)$ the different components and properties, e.g. "if $x^{(2)}$ is small and $z^{(4)}$ is large then $y^{(1)}$ is medium". The formulation of the decision problem and the corresponding considerations are the same as for one-dimensional case with

$$\varphi_u(u) = \bigwedge_{i \in \overline{1,s}} [\varphi_{ui}(1) \vee \varphi_{ui}(2) \vee ... \vee \varphi_{ui}(m)]$$

$$= \bigvee_{j_1 ... j_s \in \overline{1,m}} [\varphi_{u1}(j_1) \wedge \varphi_{u2}(j_2) \wedge ... \wedge \varphi_{us}(j_s)]$$

where s is a number of components in the vector x, and with $\varphi_z(z)$, $\varphi_y(y)$ in the analogous form. The formulas (6.8) – (6.11) have the identical form for the multidimensional case where μ_u, μ_z, μ_y, μ_{uyz}, μ_{uz} and $\mu_y(y \mid u, z)$ are the membership functions of φ_u, φ_z, φ_y, $\varphi_u \wedge \varphi_z \wedge \varphi_y$ and $\varphi_u \wedge \varphi_z \rightarrow \varphi_y$, respectively. The determinization of the fuzzy decision algorithm consists in the determination of $M(\hat{u}^{(i)})$ for the fixed z and each component of the vector u, using the membership functions $\mu_{ui}(u^{(i)}, z)$ or $\mu_{ui}(u^{(i)} \mid z)$ where

$$\mu_{ui}(u^{(i)}, z) = \max\{\mu_{ui}(1, z), \mu_{ui}(2, z), ..., \mu_{ui}(m, z)\}.$$

Example 6.1: Consider a plant with u, y, $z \in R^1$ described by the following KP:

" If u is small nonnegative and z is large but not greater than b (i.e. $b - z$ is small nonnegative) then y is medium". Then

$\varphi_u(u) = $ " u is small nonnegative",

$\varphi_z(z) = $ " z is large, not greater than b",

$\varphi_y(y) = $ " y is medium".

The membership function $w[\varphi_u \wedge \varphi_z \rightarrow \varphi_y]$ is as follows:

$$\mu_y(y \mid u, z) = -(y - d)^2 + 1 - u - (b - z)$$

for

$$0 \le u \le \frac{1}{2}, \qquad b - \frac{1}{2} \le z \le b,$$

$$-\sqrt{1 - x - (b - z)} + d \le y \le \sqrt{1 - x - (b - z)} + d$$

and $\mu_y(y \mid u, z) = 0$ otherwise.

For the membership function required by a user

$$\mu_y(y) = \begin{cases} -(y - c)^2 + 1 & \text{for} \quad c - 1 \le y \le c + 1 \\ 0 & \text{otherwise}, \end{cases}$$

one should determine the fuzzy decision algorithm in the form $\mu_u(u \mid z) = \mu_{uz}(u, z)$.

Let us assume that

$$c + 1 \le d \le c + 2.$$

Then the equation (6.11) has a unique solution which is reduced to the solution of the equation

$$-(y - c)^2 + 1 = -(y - d)^2 + 1 - u - (b - z).$$

Further considerations are the same as in Example 2.4 which is identical from the formal point of view. Consequently, we obtain the following result:

$$\mu_{uz}(u, z) = \begin{cases} -[\dfrac{(d-c)^2 + u + b - z}{2(d-c)}]^2 + 1 & \text{for} \quad u - z \le 1 - [d - (c+1)]^2 - b, \\ & \qquad 0 \le u \le \dfrac{1}{2}, \quad b - \dfrac{1}{2} \le z \le b \\ 0 & \text{otherwise} \quad . \end{cases}$$

Applying the determinization (defuzzification) we can determine $u = M(\hat{u} ; z) = \Psi(z)$, i.e. the deterministic decision algorithm in an open-loop system.

6.4 Generalisation. Soft Variables

It is worth noting the analogies between the relationships (2.17), (6.2) and (6.9), for uncertain variables, random variables and fuzzy numbers, respectively. The uncertain variables, the random variables and the fuzzy numbers may be considered as special cases of more general description of the uncertainty in the form of *soft variables* and *evaluating functions* [28, 29] which may be introduced as a tool for a unification and generalisation of decision making problems based on the uncertain knowledge representation.

Definition 6.1: A soft variable $\check{x} = \langle X, g(x) \rangle$ is defined by the set of values X and an evaluating function $g : X \to R^+$. The evaluating function satisfies the following conditions:

$$\int_X xg(x) < \infty$$

for the continuous case and

$$\sum_{i=1}^{\infty} x_i g(x_i) < \infty$$

for the discrete case, i.e. for $X = \{x_1, x_2, ..., x_\infty\}$.   □

For two soft variables $(\check{x}, \check{y})$ we can introduce the joint evaluating function $g_{xy}(x, y)$ and the conditional evaluating functions $g_x(x \mid y)$, $g_y(y \mid x)$. For example $g_x(x \mid y)$ denotes the evaluating function of $\check{x}$ for the given value y. The evaluating function may have different practical interpretation (semantics). In the random case the soft variable is the random variable described by the probability density $g(x) = f(x)$, in the case of the uncertain variable $g(x) = h(x)$ is the certainty distribution and in the case of the fuzzy description the soft variable is the fuzzy number described by the membership function $g(x) = \mu(x) = w[\varphi(x)]$ where w denotes a logic value of a given soft property $\varphi(x)$.

Let us consider the plant with the input vector $u \in U$, the output vector $y \in Y$ and the vector of disturbances $z \in Z$, and assume that (u, y, z) are values of soft variables $(\check{u}, \check{y}, \check{z})$. Denote by D_{gu}, D_{gy} and D_{gz} the sets of the evaluating functions $g_u(u)$, $g_y(y)$ and $g_z(z)$, respectively. The relation

$$R_g(g_u, g_y, g_z) \subset D_{gu} \times D_{gy} \times D_{gz}, \tag{6.12}$$

i.e. the relationship between the evaluating functions may be considered as the knowledge representation of the plant (KP). It is easy to note that R_g, g_u, g_y, g_z are generalisations of the statements: "R is a set of all possible values (u, y, z)", "D_u is a set of all possible values u" etc., introduced in Sect. 3.1. For example, if D_u is a set of all possible values u then $g_u(u) = const.$ for $u \in D_u$ and $g_u(u) = 0$ for $u \notin D_u$. If z is fixed then the relation R_g (6.12) is reduced to $R_g(g_u, g_y ; z)$. For this case let us formulate the decision problem for the required property concerning y in the form $g_y \in \overline{D}_{gy}$ where $\overline{D}_{gy} \subset D_{gy}$ is given by a user.

Decision problem: For the given $R_g(g_u, g_y ; z)$, D_{gy} and z find the largest set $D_{gu}(z)$ such that the implication

$$g_u \in D_{gu}(z) \rightarrow g_y \in D_{gy}$$

is satisfied.

Our problem is analogous to that presented in Sect. 3.2 (see 3.18) and

$$D_{gu}(z) = \{g_u \in D_{gu} : D_{gy}(g_u ; z) \subseteq D_{gy}\} \triangleq \overline{R}_g(g_u ; z) \tag{6.13}$$

where

$$D_{gy}(g_u ; z) = \{g_y \in D_{gy} : (g_u, g_y) \in R_g(g_u, g_y ; z)\}$$

and $D_{gu}(z) = \overline{R}(g_u ; z)$ denotes the set of the evaluating functions $g_u(u ; z)$ which may be considered as the knowledge of the decision making KD determined from the given knowledge of the plant KP. For the set $D_{gu}(z)$ we can determine the set S_M of the mean values $M_u(z)$ for all $g_u \in D_{gu}(z)$ and use the mean value as a final decision. Consequently, as a result based on KP we obtain the set S_Ψ of the decision algorithms $\Psi(z)$:

$$S_\Psi = \{\Psi : \Psi(z) \in S_M \quad \text{for every } z\}.$$

Denote by $\bar{g}_u(u)$ the evaluating function in the case when the set of possible values of $\overset{\vee}{u}$ is reduced to one value u. In this case $M_u = u$ and R_g is reduced to the set $D_{gy}(u, z)$ of the evaluating functions $g_y(y; u, z)$. Now we can propose the determinization of KP and KD using the mean values.

Decision problem with the determinization: For the given $D_{gy}(u, z)$, the required output value y^* and z, find the decision u such that $M_y(u, z) = y^*$.

As a result we obtain the set S_Ψ of the decision algorithms $u = \Psi(z)$ corresponding to all $g_y(y; u, z)$ in the set $D_{gy}(u, z)$. As KD we can accept $R_g(g_u, g_y, z)$ with $g_y = \bar{g}_y$ for $y = y^*$, (i.e. $\overset{\vee}{y}$ has only one possible value equal to y^*) which is reduced to the set $D_{gu}(z)$ of the evaluating functions $g_u(u; z)$. The determinization of KD gives the set S_{Ψ_d} of the decision algorithms based on KD:

$$u_d = M_u(z) \overset{\Delta}{=} \Psi_d(z)$$

where $M_u(z)$ is the mean value for $g_u(u; z)$ belonging to the set $D_{gu}(z)$. The solutions based on KP and KD may be not equivalent, i.e. in general $S_\Psi \neq S_{\Psi_d}$. This was shown for the uncertain variables, that is in the case where the evaluating functions are the certainty distributions (see Example 2.3 in Sect. 2.3 for the parametric case).

The relation R_g may have the form of a function (a one-to-one mapping):

$$g_u = T_u(g_y; z) \quad \text{or} \quad g_y = T_y(g_u; z) \tag{6.14}$$

i.e.

$$T_u : D_{gy} \rightarrow D_{gu} \quad \text{or} \quad T_y : D_{gu} \rightarrow D_{gy}.$$

Then (6.13) is reduced to

$$D_{gu}(z) = \{g_u = T_u(g_y; z)\}$$

or

$$D_{gu}(z) = \{g_u : T_y(g_u; z) = g_y\}. \tag{6.15}$$

In particular, if $g_y = \bar{g}_y(y)$ or $g_u = \bar{g}_u(u)$ then according to (6.14)

$$g_u = g_u(u;y,z) \quad \text{or} \quad g_y = g_y(y;u,z),$$

respectively.

The relationships between the general formulation using soft variables and the respective formulations with uncertain, random and fuzzy variables may be shown directly for the fixed z. Then in the place of (2.17), (6.2) and (6.9) we have

$$h_y(y) = \max_{u \in U, z \in Z} \min\{h_u(u;z), h_y(y \mid u;z)\}, \tag{6.16}$$

$$f_y(y) = \int_U \int_Z f_u(u;z) f_y(y \mid u;z) du dz, \tag{6.17}$$

$$\mu_y(y) = \max_{u \in U, z \in Z} \min\{\mu_u(u;z), \mu_y(y \mid u;z)\}, \tag{6.18}$$

respectively. In this formulations $h_u(u;z)$ and $h_y(y \mid u;z)$ denote the certainty distributions for the fixed value z; $f_u(u;z)$, $f_y(y \mid u;z)$ are the probability densities and $\mu_u(u;z)$, $\mu_y(y \mid u;z)$ are the membership functions for the fixed value z. In the first case we assume that the soft variables $(\breve{u}, \breve{y})$ are the uncertain variables, the evaluating functions take the form of the certainty distributions

$$g_y(y) = h_y(y), \quad g_y(y;u,z) = h_y(y \mid u;z) \tag{6.19}$$

and the function T_y is determined by (6.16).

The knowledge of the plant KP is then reduced to $g_y(y;u,z) = h_y(y \mid u;z)$. For the required distribution $h_y(y)$, according to (6.15), the result of the decision problem based on KP is the set $D_{gu}(z)$ of the distributions $h_u(u;z)$ satisfying the equation (6.16).

In the second case

$$g_y(y) = f_y(y), \quad g_y(y;u,z) = f_y(y \mid u;z),$$

the function T_y is determined by (6.17) and the result of the decision problem based on KP $= < f_y(y \mid u;z) >$ is the set $D_{gu}(z)$ of the probability densities $f_u(u;z)$ satisfying the equation (6.17) for the required density $f_y(y)$.

In the third case

$$g_y(y) = \mu_y(y), \qquad g_y(y;u,z) = \mu_y(y \mid u;z),$$

T_y is determined by (6.18) and as the result of the decision problem based on $KP = < \mu_y(y \mid u;z) >$ we obtain the set $D_{gu}(z)$ of the membership functions $\mu_u(u;z)$.

An evaluating function $g_u(u;z)$ chosen from the set $D_{gu}(z)$ may be called a *soft decision algorithm* in an open-loop decision system. The uncertain decision algorithm $h_u(u;z)$, the random decision algorithm $f_u(u;z)$ and the fuzzy decision algorithm $\mu_u(u;z)$ may be considered as special cases of the soft decision algorithm. If z is assumed to be a value of a soft variable $\overset{\vee}{z}$ (in particular, a value of an uncertain variable $\overline{z}$, a random variable $\tilde{z}$ or a fuzzy number $\hat{z}$) then we can introduce the evaluating function $g_z(z)$ (in particular, $h_z(z)$, $f_z(z)$ or $\mu_z(z)$). Instead of (6.16), (6.17), (6.18) we have the formulations (2.17), (6.2) and (6.9) which may be generalised by introducing soft variables and the conditional evaluating functions.

7 Special and Related Problems

7.1 Pattern Recognition

Let an object to be recognized or classified be characterized by a vector of features $u \in U$ which may be observed, and the index of a class j to which the object belongs; $j \in \{1, 2, ..., M\} \overset{\Delta}{=} J$, M is a number of the classes. The set of the objects may be described by a relational knowledge representation $R(u, j) \in U \times J$ which is reduced to the sequence of sets

$$D_u(j) \subset U, \qquad j = 1, 2, ..., M,$$

i.e.

$$D_u(j) = \{u \in U : (u, j) \in R(u, j)\}.$$

Assume that as a result of the observation it is known that $u \in D_u \subset U$. The recognition problem may consist in finding the set of all possible indexes j, i.e. the set of all possible classes to which the object may belong [32, 49].

Recognition problem: For the given sequence $D_u(j)$, $j \in \overline{1, M}$ and the result of observation D_u find the smallest set $D_j \subset J$ for which the implication

$$u \in D_u \rightarrow j \in D_j$$

is satisfied.

This is the specific analysis problem for the relational plant (see Sect. 3.2) and

$$D_j = \{j \in J : D_u \cap D_u(j) \neq \emptyset\}$$

where $\emptyset$ denotes an empty set. In particular, if $D_u = \{u\}$, i.e. we obtain the exact result of the measurement then

$$D_j = \{j \in J : u \in D_u(j)\}.$$

Now let us assume that the knowledge representation contains a vector of un-known parameters $x \in X$ and x is assumed to be a value of an uncertain vari-able $\bar{x}$ described by a certainty distribution $h_x(x)$ given by an expert. The recog-nition problem is now formulated as a specific analysis problem (version I) considered in Sect. 3.4.

Recognition problem for uncertain parameters: For the given sequence $D_u(j;x)$, $h_x(x)$, D_u and the set $\hat{D}_j \subset J$ given by a user one should find the certainty index that the set $\hat{D}_j$ belongs to the set of all possible classes

$$D_j(x) = \{ j \in J : D_u \cap D_u(j;x) \neq \varnothing \}. \tag{7.1}$$

It is easy to see that

$$v[\hat{D}_j \tilde{\subseteq} D_j(\bar{x})] = v[\bar{x} \tilde{\in} D_x(\hat{D}_j)] \tag{7.2}$$

where

$$D_x(\hat{D}_j) = \{ x \in X : \hat{D}_j \subseteq D_j(x) \}. \tag{7.3}$$

Then

$$v[\hat{D}_j \tilde{\subseteq} D_j(\bar{x})] = \max_{x \in D_x(\hat{D}_j)} h_x(x). \tag{7.4}$$

In particular, for $\hat{D}_j = \{j\}$ one can formulate the optimization problem consisting in the determination of a class j maximizing the certainty index that j belongs to the set of all possible classes.

Optimal recognition problem: For the given sequence $D_u(j;x)$, $h_x(x)$ and D_u one should find j^* maximizing

$$v[j \tilde{\in} D_j(\bar{x})] \overset{\Delta}{=} v(j).$$

Using (7.2), (7.3) and (7.4) for $\hat{D}_j = \{j\}$ we obtain

$$v(j) = v[\bar{x} \tilde{\in} D_x(j)] = \max_{x \in D_x(j)} h_x(x) \tag{7.5}$$

where

$$D_x(j) = \{ x \in X : j \in D_j(x) \} \tag{7.6}$$

and $D_j(x)$ is determined by (7.1). Then

$$j^* = \arg\max_j v(j) = \arg\max_j \max_{x \in D_x(j)} h_x(x). \tag{7.7}$$

Assume that the different unknown parameters are separated in the different sets, i.e. the knowledge representation is described by the sets $D_u(j; x_j)$ where $x_j \in X_j$ are subvectors of x, different for the different j. Assume also that $\bar{x}_j$ and $\bar{x}_i$ are independent uncertain variables for $i \neq j$ and $\bar{x}_j$ is described by the certainty distribution $h_{xj}(x_j)$. In this case, according to (7.1)

$$j \in D_j(x) \Leftrightarrow D_u \cap D_u(j; x_j) \neq \varnothing.$$

Then

$$v(j) = v[\bigvee_{u \in D_u} u \, \tilde{\in} \, D_u(j; \bar{x}_j)] = v[\bar{x}_j \, \tilde{\in} \, D_{xj}(j)] \qquad (7.8)$$

where

$$D_{xj}(j) = \{x_j \in X_j : \bigvee_{u \in D_u} u \in D_u(j; x_j)\}. \qquad (7.9)$$

Finally

$$j^* = \arg\max_{j} \max_{x_j \in D_{xj}(j)} h_{xj}(x_j). \qquad (7.10)$$

In particular, for $D_u = \{u\}$ (7.1), (7.8) and (7.9) become

$$D_j(x) = \{j \in J : u \in D_u(j; x)\}$$

$$v(j) = v[u \, \tilde{\in} \, D_u(j; \bar{x}_j)] = v[\bar{x}_j \, \tilde{\in} \, D_{xj}(j)] = \max_{x_j \in D_{xj}(j)} h_{xj}(x_j), \quad (7.11)$$

$$D_{xj}(j) = \{x_j \in X_j : u \in D_u(j; x_j)\}. \qquad (7.12)$$

The procedure of finding j^* based on the knowledge representation $< D_u(j; x),\ j \in \overline{1, M}\,;\ h_x(x) >$ or the block scheme of the corresponding recognition system is illustrated in Fig. 7.1. The solution may be not unique, i.e. $v(j)$ may take the maximum value for the different j^*. The result $v(j) = 0$ for each $j \in J$ means that the result of the observation $u \in D_u$ is not possible or there is a contradiction between the result of the observation and the knowledge representation given by an expert.

If $\bar{x}$ is considered as C-uncertain variable then

$$j_c^* = \arg\max_{j} v_c(j)$$

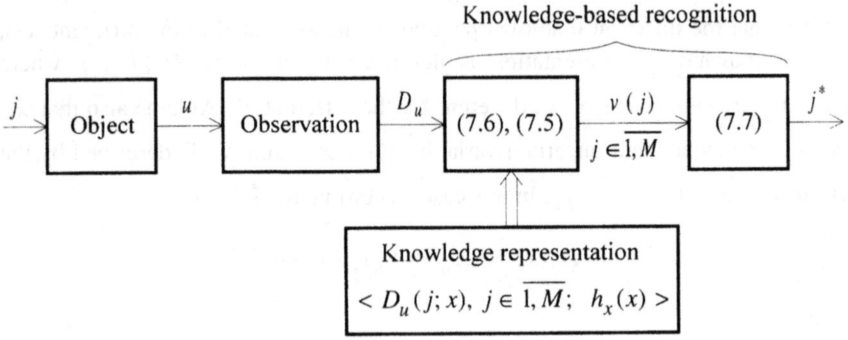

Fig. 7.1. Block scheme of recognition system

where

$$v_c(j) = \frac{1}{2}\{v[\bar{x} \tilde{\in} D_x(j)] + 1 - v[\bar{x} \tilde{\in} \overline{D}_x(j)]\},$$

$\overline{D}_x(j) = X - D_x(j)$. Finally

$$v_c(j) = \frac{1}{2}[\max_{x \in D_x(j)} h_x(x) + 1 - \max_{x \in \overline{D}_x(j)} h_x(x)]. \qquad (7.13)$$

The certainty indexes $v_c(j)$ corresponding to (7.8) and (7.11) have the analogous form.

Example 7.1: Let $u, x_j \in R^1$, the sets $D_u(j; x_j)$ be described by the inequalities

$$x_j \leq u \leq 2x_j, \quad j = 1, 2, ..., M$$

and the certainty distributions $h_{xj}(x_j)$ have a parabolic form for each j (Fig. 7.2):

$$h_{xj}(x_j) = \begin{cases} -(x_j - d_j)^2 + 1 & \text{for } d_j - 1 \leq x_j \leq d_j + 1 \\ 0 & \text{otherwise} \end{cases}$$

where $d_j > 1$.

In this case the sets (7.12) for the given u are described by the inequality

$$\frac{u}{2} \leq x_j \leq u.$$

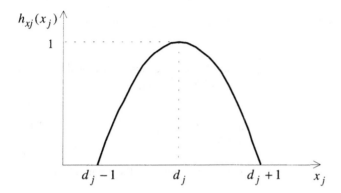

Fig. 7.2. Parabolic certainty distribution

Applying (7.11) one obtains $v(j)$ as a function of d_j illustrated on Fig. 7.3:

$$v(j) = \begin{cases} 0 & \text{for} & d_j \leq \frac{u}{2} - 1 \\ -(\frac{u}{2} - d_j)^2 + 1 & \text{for} & \frac{u}{2} - 1 \leq d_j \leq \frac{u}{2} \\ 1 & \text{for} & \frac{u}{2} \leq d_j \leq u \\ -(u - d_j)^2 + 1 & \text{for} & u \leq d_j \leq u + 1 \\ 0 & \text{for} & d_j \geq u + 1. \end{cases}$$

For example, for $M = 3$, $u = 5$, $d_1 = 2$, $d_2 = 5.2$, $d_3 = 6$ we obtain
$v(1) = 0.75$, $v(2) = 0.96$ and $v(3) = 0$. Then $j^* = 2$, which means that for
$u = 5$ the certainty index that $j = 2$ belongs to the set of the possible classes has

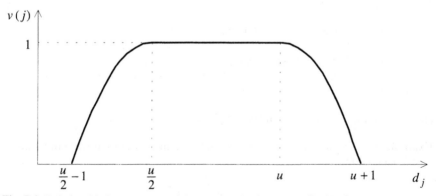

Fig. 7.3. Relationship between v and the parameter of certainty distribution

the maximum value equal to 0.96. For $d_1, d_2, d_3 \in [\frac{u}{2}, u]$ one obtains $j^* = 1$ or 2 or 3 and $v(j^*) = 1$.

Let us consider $\bar{x}$ as a C-uncertain variable for the same numerical data. To obtain $v_c(j)$ according to (7.13) it is necessary to determine

$$v[\bar{x}_j \tilde{\in} \bar{D}_{xj}(j)] = \max_{x_j \in \bar{D}_{xj}(j)} h_{xj}(x_j) \overset{\Delta}{=} v_n(j). \qquad (7.14)$$

In our case the set $\bar{D}_{xj}(j) = X_j - D_{xj}(j)$ is determined by the inequalities

$$x_j < \frac{u}{2} \qquad \text{or} \qquad x_j > u.$$

Using (7.14) we obtain $v_n(1) = v_n(2) = v_n(3) = 1$. Then

$$v_c(j) = \frac{1}{2}[v(j) + 1 - v_n(j)] = \frac{1}{2}v(j), \qquad (7.15)$$

i.e. $v_c(1) = 0.375$, $v_c(2) = 0.48$, $v_c(3) = 0$ and $j_c^* = 2$ with the certainty index $v_c(j^*) = 0.48$.

For $d_1 = 3$, $d_2 = 3.2$, $d_3 = 4$ we obtain $v(1) = v(2) = v(3) = 1$ and

$$v_n(1) = -(2.5 - 3)^2 + 1 = 0.75,$$
$$v_n(2) = -(2.5 - 3.2)^2 + 1 = 0.51,$$
$$v_n(3) = 0.$$

Then

$$v_c(1) = \frac{1}{2}(1 + 1 - 0.75) = 0.625,$$
$$v_c(2) = \frac{1}{2}(1 + 1 - 0.51) = 0.745,$$
$$v_c(3) = 1$$

and $j_c^* = 3$ with the certainty index $v_c(j_c^*) = 1$.

Example 7.2: Assumed that in the Example 7.1 the certainty distributions have an exponential form:

$$h_{xj}(x_j) = e^{-(x_j - d_j)^2}.$$

Applying (7.11) one obtains $v(j)$ as a function of d_j:

$$v(j) = \begin{cases} e^{-(\frac{u}{2}-d_j)^2} & \text{for} & d_j \leq \frac{u}{2} \\ 1 & \text{for} & \frac{u}{2} \leq d_j \leq u \\ e^{-(u-d_j)^2} & \text{for} & d_j \geq u . \end{cases}$$

For $M = 3$, $u = 5$, $d_1 = 2$, $d_2 = 5.2$, $d_3 = 6$ we obtain

$$v(1) = e^{-0.25}, \quad v(2) = e^{-0.4}, \quad v(3) = e^{-1}.$$

Then $j^* = 2$ with the certainty index $v(j^*) = e^{-0.4} = 0.67$. For $d_1, d_2, d_3 \in [\frac{u}{2}, u]$ one obtains $j^* = 1$ or 2 or 3 and $v(j^*) = 1$.

Now let us consider $\bar{x}$ as an C-uncertain variable. Using (7.14) we obtain

$$v_n(j) = \begin{cases} 1 & \text{for} & d_j \leq \frac{u}{2} \\ e^{-(\frac{u}{2}-d_j)^2} & \text{for} & \frac{u}{2} \leq d_j \leq \frac{3}{4}u \\ e^{-(u-d_j)^2} & \text{for} & \frac{3}{4}u \leq d_j \leq u \\ 1 & \text{for} & d_j \geq u . \end{cases}$$

Then the formula

$$v_c(j) = \frac{1}{2}[v(j) + 1 - v_n(j)]$$

gives the following results:

$$v_c(j) = \begin{cases} \frac{1}{2}e^{-(\frac{u}{2}-d_j)^2} & \text{for} & d_j \leq \frac{u}{2} \\ 1 - \frac{1}{2}e^{-(\frac{u}{2}-d_j)^2} & \text{for} & \frac{u}{2} \leq d_j \leq \frac{3}{4}u \\ 1 - \frac{1}{2}e^{-(u-d_j)^2} & \text{for} & \frac{3}{4}u \leq d_j \leq u \\ \frac{1}{2}e^{-(u-d_j)^2} & \text{for} & d_j \geq u . \end{cases}$$

Substituting the numerical data $u = 5$, $d_1 = 2$, $d_2 = 5.2$, $d_3 = 6$ one obtains

$$v_c(1) = \frac{1}{2}e^{-0.25}, \quad v_c(2) = \frac{1}{2}e^{-0.04}, \quad v_c(3) = \frac{1}{2}e^{-1}.$$

Then $j_c^* = j^* = 2$ with the certainty index $v_c(j_c^*) = \frac{1}{2}e^{-0.4} = 0.335$. The results for $d_1 = 3$, $d_2 = 3.2$ and $d_3 = 4$ are as follows:

$$v_c(1) = 1 - \frac{1}{2}e^{-0.25}, \qquad v_c(2) = 1 - \frac{1}{2}e^{-0.49}, \qquad v_c(3) = 1 - \frac{1}{2}e^{-1}$$

and $j_c^* = 3$ with the certainty index $v_c(j_c^*) = 1 - \frac{1}{2}e^{-1} = 0.816$.

In this particular case the results j^* and j_c^* are the same for the different forms of the certainty distribution (see Example 7.1).

7.2 Control of the Complex of Operations

The uncertain variables may be applied to a special case of the control of the complex of parallel operations containing unknown parameters in the relational knowledge representation. The control consists in a proper distribution of a given size of a task taking into account the execution time of the whole complex. It may mean the distribution of a raw material in the case of a manufacturing process or a load distribution in a group of parallel computers. In the deterministic case where the operations are described by functions determining the relationship between the execution time and the size of the task, the optimization problem consisting in the determination of the distribution minimizing the execution time of the complex may be formulated and solved (see e.g. [7]). In the case of the relational knowledge representation with uncertain parameters the problem consists in the determination of the distribution maximizing the certainty index that the requirement given by a user is satisfied [24, 30, 31]. This is a specific form of the decision problem described in Sect. 3.5.

Let us consider the operations described by the relations

$$R_i(u_i, T_i ; x_i) \subset R^+ \times R^+ , \qquad i = 1, 2, ..., k \qquad (7.16)$$

where u_i is the size of the task, T_i denotes the execution time and x_i is an unknown parameter which is assumed to be a value of an uncertain variable $\bar{x}_i$ with the certainty distribution $h_{xi}(x_i)$ given by an expert. From (7.16) we obtain the set of the possible values T_i for the given value u_i:

$$D_{T,i}(u_i ; x_i) = \{T_i : (u_i, T_i) \in R_i(u_i, T_i ; x_i)\} .$$

The complex of operations is considered as a plant with the input $u = (u_1, u_2, ..., u_k)$, the output $y = T$ where T is the execution time of the whole complex:

$$T = \max\{T_1, T_2, ..., T_k\}, \tag{7.17}$$

and the requirement $T \in [0, \alpha]$, i.e. $T \leq \alpha$ where α is a number given by a user.
Decision problem: For the given R_i, $h_{xi}(x_i)$ $(i = 1, 2, ..., k)$ and α find the distribution $\hat{u} = (\hat{u}_1, \hat{u}_2, ..., \hat{u}_k)$ maximizing the certainty index $v(u)$ that the approximate set of the possible outputs $y \in T$ belongs to the interval $[0, \alpha]$, subject to constraints

$$u_1 + u_2 + ... + u_k = \overline{U}, \quad \bigwedge_{i \in \overline{1, k}} u_i \geq 0 \tag{7.18}$$

where $\overline{U}$ is the global size of the task to be distributed.

From (7.17) it is easy to note that the requirement $T \leq \alpha$ is equivalent to the requirement

$$(T_1 \in [0, \alpha]) \wedge (T_2 \in [0, \alpha]) \wedge ... \wedge (T_k \in [0, \alpha]).$$

Then

$$\hat{u} = \arg\max_{u \in U} v(u) = \arg\max_{u \in U} \min_i v_i(u_i)$$

where U is determined by the constraints (7.18) and

$$v_i(u_i) = v\{D_{T,i}(u_i; \overline{x}_i) \subseteq [0, \alpha]\}. \tag{7.19}$$

Consider as a special case the relations (7.16) described by the inequalities

$$T_i \leq x_i u_i, \quad x_i > 0, \quad i = 1, 2, ..., k. \tag{7.20}$$

The inequality (7.20) determines the set of possible values of the execution time in i-th operation for the fixed value of the size of the task u_i, e.g. the set of possible values of the processing time for the amount of a raw material equal to u_i, in the case of a manufacturing process. Then the certainty index (7.19) is reduced to the following form

$$v_i(u_i) = v(\overline{T}_i \lesssim \alpha) = v[\overline{x}_i \tilde{\in} D_{xi}(u_i)] = \max_{x_i \in D_{xi}(u_i)} h_{xi}(x_i) \tag{7.21}$$

where $D_{xi}(u_i)$ is described by the inequality

$$x_i \leq \alpha u_i^{-1}.$$

Example 7.3: Let $h_{xi}(x_i)$ have a triangular form (Fig. 7.4). Then, using (7.21) it is easy to obtain

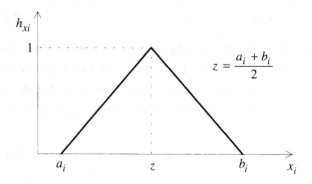

Fig. 7.4. Triangular certainty distribution

$$v_i(u_i) = \begin{cases} 1 & \text{for} & u_i \le 2\alpha(a_i + b_i)^{-1} \\ 0 & \text{for} & u_i \ge \alpha a_i^{-1} \\ -A_i u_i + B_i & & \text{otherwise} \end{cases}$$

where

$$A_i = \frac{a_i(a_i + b_i)}{\alpha(b_i - a_i)}, \qquad B_i = \frac{a_i + b_i}{b_i - a_i}.$$

For $k = 2$ the decision $\hat{u}_1$ may be found by solving the equation

$$v_1(u_1) = v_2(\overline{U} - u_1).$$

The result is as follows:

1.   For $\overline{U} \ge \alpha(\dfrac{1}{a_1} + \dfrac{1}{a_2})$

   $v(u) = 0$ for any u_1 which means that α is too small to satisfy the requirement.

2.   For $\overline{U} \le 2\alpha[\dfrac{1}{a_1 + b_1} + \dfrac{1}{a_2 + b_2}]$

   the optimal decision $\hat{u}_1$ is any value from the interval

$$[\overline{U} - \frac{2\alpha}{a_2 + b_2}, \quad \frac{2\alpha}{a_1 + b_1}]$$

   and

$$\max_u v(u) \triangleq v^* = 1.$$

3.   Otherwise,

$$\hat{u}_1 = (B_1 - B_2 + A_2\overline{U})(A_1 + A_2)^{-1} \qquad (7.22)$$

and $v^* = B_1 - A_1 \hat{u}_1$.

For example, if $\overline{U} = 2$, $\alpha = 2$, $a_1 = 1$, $b_1 = 3$, $a_2 = 2$ and $b_2 = 4$ then $\hat{u}_1$ is determined from (7.22) and $\hat{u}_1 = 1.25$, $\hat{u}_2 = 0.75$. For this distribution the requirement $T \le \alpha$ is approximately satisfied with the certainty index $v^* = 0.75$.

7.3   Descriptive and Prescriptive Approaches

In the analysis and design of knowledge-based uncertain systems it may be important to investigate a relation between two concepts concerning two different subjects of the knowledge given by an expert [29]. In the *descriptive approach* an expert gives the knowledge of the plant KP, and the knowledge of the decision making KD is obtained from KP for the given requirement. This approach is widely used in the traditional decision and control theory. The deterministic decision algorithm may be obtained via the determinization of KP or the determinization of KD based on KP. Such a situation for the formulation using uncertain variables is illustrated in Figs. 2.2, 2.3, 2.4, 3.5 and 3.6. In the *prescriptive approach* the knowledge of the decision making $\overline{KD}$ is given directly by an expert. This approach is used in the design of fuzzy controllers where the deterministic control algorithm is obtained via the defuzzification of the knowledge of the control given by an expert. The descriptive approach to the decision making based on the fuzzy description may be found in [39].

Generally speaking, the descriptive and prescriptive approaches may be called *equivalent* if the deterministic decision algorithms based on KP and $\overline{KD}$ are the same. Different particular cases considered in the previous chapters may be illustrated in Figs. 7.5 and 7.6 for two different concepts of the determinization. Fig. 7.7 illustrates the prescriptive approach. In the first version (Fig. 7.5) the approaches are equivalent if $\Psi(z) = \overline{\Psi}_d(z)$ for every z. In the second version (Fig. 7.6) the approaches are equivalent if $KD = \overline{KD}$. Then $\Psi(z) = \overline{\Psi}_d(z)$ for every z.

Let us consider more precisely version I of the decision problem described in Sect. 2.3. An expert formulates $KP = \langle \Phi, h_x \rangle$ (the descriptive approach) or $\overline{KD} = \langle \overline{\Phi}_d, h_x \rangle$ (the prescriptive approach). In the first version of the determinization illustrated in Fig. 2.2 the approaches are equivalent if $\Psi_a(z) = \overline{\Psi}_{ad}(z)$ for every z, where $\Psi_a(z)$ is determined by (2.8) and $\overline{\Psi}_{ad}(z)$ is determined by (2.13) with $\overline{\Phi}_d(z, x)$ instead of $\Phi_d(z, x)$ obtained as a solution of the equation

$$\Phi(u, z, x) = \hat{y}. \tag{7.23}$$

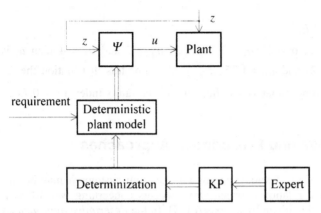

Fig. 7.5. Illustration of descriptive approach – the first version

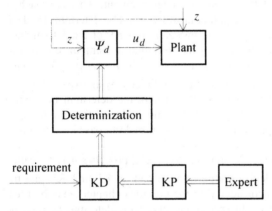

Fig. 7.6. Illustration of descriptive approach – the second version

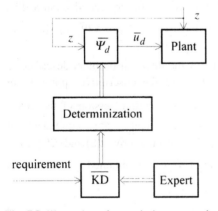

Fig. 7.7. Illustration of prescriptive approach

In the second version of the determinization illustrated in Fig. 2.3 the approaches are equivalent if the solution of the equation (7.23) with respect to u has the form $\overline{\varPhi}_d(z, x)$, i.e.

$$\varPhi[\overline{\varPhi}_d(z, x), z, x] = \hat{y}.$$

For the nonparametric problem described in Sect. 2.4 only the second version of the determinization illustrated in Fig. 2.4 may be applied. If we accept (2.20) as a solution of the equation (2.16) then $\text{KP} = < h_y(y \mid u, z) >$ and $\overline{\text{KD}} = < \overline{h}_u(u \mid z) >$ are equivalent for the given required distribution $h_y(y)$ if $\overline{h}_u(u \mid z) = \overline{h}_{uz}(u, z)$ satisfies the equation (2.18). The similar formulation of the equivalency may be given for the random and the fuzzy descriptions presented in Sects. 6.2 and 6.3, respectively.

The generalisation for the soft variables and evaluating functions described in Sect. 6.4 may be formulated as a principle of equivalency.

Principle of equivalency: If the knowledge of the decision making $\overline{\text{KD}}$ given by an expert has a form of the set of evaluating functions $\overline{D}_{gu}(z)$ and $\overline{D}_{gu}(z) \subseteq D_{gu}(z)$ where $D_{gu}(z)$ is determined by (6.13), then $\overline{S}_\psi \subseteq S_\psi$ where $\overline{S}_\psi$ is the set of the decision algorithms corresponding to $\overline{D}_{gu}(z)$. In particular, if an expert gives one evaluating function $\overline{g}_{gu}(u; z)$, i.e. $\overline{D}_{gu}(z) = \{\overline{g}_{gu}(u; z)\}$ and $\overline{g}_{gu}(u; z) \in D_{gu}(z)$ then the decision algorithm based on the knowledge of the decision making given by an expert is equivalent to one of the decision algorithms based on the knowledge of the plant.

7.4 Complex Uncertain System

As an example of a complex uncertain system let us consider two-level system presented in Fig. 7.8, described by a relational knowledge representation with uncertain parameters where $u_i \in U_i$, $y_i \in Y_i$, $z_i \in Z_i$, $y \in Y$ [22, 30]. For example, it may be a production system containing k parallel operations (production units) in which y_i is a vector of variables characterizing the product (e.g. the amounts of some components), u_i is a vector of variables characterizing the raw material which are accepted as the control variables and z_i is a vector of disturbances which are measured. The block P may denote an additional production unit or an evaluation of a vector y of global variables characterizing the system as a whole.

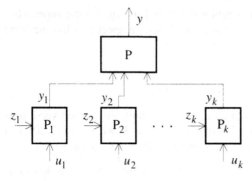

Fig. 7.8. Example of complex system

Assume that the system is described by a relational knowledge representation which has a form of a set of relations:

$$\left.\begin{aligned} R_i(u_i, y_i, z_i ; x_i) &\subset U_i \times Y_i \times Z_i, \quad i \in \overline{1, k} \\ R_y(\overline{y}, y ; x_{k+1}) &\subset \overline{Y} \times Y \end{aligned}\right\} \tag{7.24}$$

where $x_i \in X_i$ $(i = 1, 2, ..., k + 1)$ are vectors of parameters,

$$\overline{y} = (y_1, y_2, ..., y_k) \in \overline{Y} .$$

Each relation may be presented as a set of inequalities and/or equalities concerning the components of the respective vectors. The unknown parameters x_i are assumed to be values of uncertain variables described by certainty distributions $h_{xi}(x_i)$ given by an expert. The relations (7.24) may be reduced to one relation

$$R(u, y, z; x) \in U \times Y \times Z$$

where

$$u = (u_1, u_2, ..., u_k) \subset U, \qquad z = (z_1, z_2, ..., z_k) \subset Z,$$

$$x = (x_1, x_2, ..., x_{k+1}) \subset X .$$

Now the decision problem may be formulated directly for the system as a whole, i.e. for the plant with the input u, the output y, the disturbance z and the uncertain vector parameter x. The formulation of the solution of the decision problem with the requirement $y \in D_y$ given by a user have been described in Sect. 3.5. If the uncertain variables $x_1, x_2, ..., x_{k+1}$ are independent then

$$h_x(x) = \min\{h_{x1}(x_1), h_{x2}(x_2), ..., h_{x,k+1}(x_{k+1})\} .$$

The direct solution of the decision problem for the system as a whole may be very complicated and it may be reasonable to apply a *decomposition*, i.e. to decompose our decision problem into separate subproblems for the block P and the blocks P_i.

1. The decision problem for the block P: For the given $R_y(\bar{y}, y; x_{k+1})$, $h_{x,k+1}(x_{k+1})$ and D_y find $\bar{y}$ maximizing the certainty index

$$v[D_y(\bar{y}; x_{k+1}) \tilde{\subseteq} D_y]$$

where

$$D_y(\bar{y}; x_{k+1}) = \{y \in Y : (\bar{y}, y) \in R_y(\bar{y}, y; x_{k+1})\}.$$

2. The decision problem for the blocks P_i $(i \in \overline{1,k})$: For the given $R_i(u_i, y_i, z_i; x_i)$, $h_{xi}(x_i)$ and D_{yi} find $\hat{x}_i$ maximizing the certainty index

$$v[D_{yi}(u_i, z_i; x_i) \tilde{\subseteq} D_{yi}]$$

where

$$D_{yi}(u_i, z_i; x_i) = \{y_i \in Y_i : (u_i, y_i, z_i) \in R_i(u_i, y_i, z_i; x_i)\}.$$

The decision problem for the block P_i with the given D_{yi} is then the same as the problem for the system as a whole with the given D_y. The sets D_{yi} are such that

$$D_{y1} \times D_{y2} \times ... \times D_{yk} \subseteq D_{\bar{y}} \qquad (7.25)$$

where $D_{\bar{y}}$ is the set of the solutions $\bar{y}$ of the decision problem for the block P.

In general, the results of the decomposition are not unique (the condition (7.25) may be satisfied by different sets D_{yi}) and differ from the results obtained from the direct approach for the system as a whole.

If there is no unknown parameter in the block P then the decomposition may have the following form:

1. The decision problem for the block P: For the given $R_y(\bar{y}, y)$ and D_y find the largest set $\hat{D}_{\bar{y}}$ such that the implication

$$\bar{y} \in \hat{D}_{\bar{y}} \rightarrow y \in D_y$$

is satisfied. This is a decision problem for the relational plant described in Sect. 3.2.

2. The decision problem for the block P_i is the same as in the previous formulation, with $\hat{D}_{\bar{y}}$ instead of $D_{\bar{y}}$.

7.5 Learning System

When the information on the unknown parameters x in the form of the certainty distribution $h_x(x)$ is not given, a learning process consisting in *step by step* knowledge validation and updating based on results of current observations may be applied [10, 18, 19, 23, 24, 25]. The results of the successive estimation of the unknown parameters may be used in the current determination of the decisions in an open-loop or closed-loop learning decision making system. This approach may be considered as an extension of the known idea of adaptation via identification for the plants described by traditional mathematical models (see e.g. [8]). Let us explain this idea for the plant described by the relation $R(u, y; x)$, considered in Chap. 3. For the given D_y one may determine the largest set of the decisions $D_u(x) \subset U$ such that the implication

$$u \in D_u(x) \rightarrow y \in D_y$$

is satisfied. The learning process may concern the knowledge of the plant $R(u, y; x)$ or directly the knowledge of the decision making $D_u(x)$. Let us consider the second case and assume that $D_u(x)$ is a continuous and closed domain in U, the parameter x has the value $x = \bar{x}$ and $\bar{x}$ is unknown. In each step of the learning process one should prove if the current observation "belongs" to the knowledge representation determined to this step (*knowledge validation*) and if not – one should modify the estimation of the parameters in the knowledge representation (*knowledge updating*). The successive estimations will be used in the determination of the decision based on the current knowledge in the learning system.

When the parameter $\bar{x}$ is unknown then for the fixed value u it is not known if u is a correct decision, i.e. if $u \in D_u(\bar{x})$ and consequently $y \in D_y$. Our problem may be considered as a classification problem with two classes. The point u should be classified to class $j = 1$ if $u \in D_u(\bar{x})$ and to class $j = 2$ if $u \notin D_u(\bar{x})$. Assume that we can use the learning sequence

$$(u_1, j_1), (u_2, j_2), ..., (u_n, j_n) \triangleq S_n$$

where $j_i \in \{1, 2\}$ are the results of the correct classification given by an external trainer or obtained by testing the property $y_i \in D_y$ at the output of the plant. Let us denote by $\bar{u}_i$ the subsequence for which $j_i = 1$, i.e. $\bar{u}_i \in D_u(\bar{x})$ and by $\hat{u}_i$ the subsequence for which $j_i = 2$, and introduce the following sets in X:

$$\overline{D}_x(n) = \{x \in X : \bar{u}_i \in D_u(x) \text{ for every } \bar{u}_i \text{ in } S_n\},$$

$$\hat{D}_x(n) = \{x \in X : \hat{u}_i \in U - D_u(x) \text{ for every } \hat{u}_i \text{ in } S_n\}.$$

It is easy to see that $\overline{D}_x$ and $\hat{D}_x$ are closed sets in X. The set

$$\overline{D}_x(n) \cap \hat{D}_x(n) \stackrel{\Delta}{=} \Delta_x(n)$$

is proposed here as the estimation of $\bar{x}$. For example, if $\bar{x} \in R^1$ and $D_u(\bar{x})$ is described by the inequality $u^T u \le \bar{x}^2$ then

$$\overline{D}_x(n) = [x_{\min,n}, \infty), \qquad \hat{D}_x(n) = [0, x_{\max,n})$$

$$\Delta_x(n) = [x_{\min,n}, x_{\max,n})$$

where

$$x_{\min,n}^2 = \max_i \bar{u}_i^T \bar{u}_i, \quad x_{\max,n}^2 = \min_i \hat{u}_i^T \hat{u}_i.$$

The determination of $\Delta_x(n)$ may be presented in the form of the following recursive algorithm:
If $j_n = 1$ ($u_n = \bar{u}_n$)
1. **Knowledge validation** for $\bar{u}_n$. Prove if

$$\bigwedge_{x \in \overline{D}_x(n-1)} [u_n \in D_u(x)].$$

If yes then $\overline{D}_x(n) = \overline{D}_x(n-1)$. If not then one should determine the new $\overline{D}_x(n)$, i.e. update the knowledge.
2. **Knowledge updating** for $\bar{u}_n$

$$\overline{D}_x(n) = \{x \in \overline{D}_x(n-1) : u_n \in D_u(x)\}.$$

Put $\hat{D}_x(n) = \hat{D}_x(n-1)$.
If $j_n = 2$ ($u_n = \hat{u}_n$)
3. **Knowledge validation** for $\hat{u}_n$. Prove if

$$\bigwedge_{x \in \hat{D}_x(n-1)} [u_n \in U - D_u(x)].$$

If yes then $\hat{D}_x(n) = \hat{D}_x(n-1)$. If not then one should determine the new $\hat{D}_x(n)$, i.e. update the knowledge.

4. **Knowledge updating** for $\hat{u}_n$

$$\hat{D}_x(n) = \{x \in \hat{D}_x(n-1) : u_n \in U - D_u(x)\}.$$

Put $\overline{D}_x(n) = \overline{D}_x(n-1)$ and $\Delta_x(n) = \overline{D}_x(n) \cap \hat{D}_x(n)$.

For $n = 1$, if $u_1 = \overline{u}_1$ determine

$$\overline{D}_x(1) = \{x \in X : u_1 \in D_u(x)\},$$

if $u_1 = \hat{u}_1$ determine

$$\hat{D}_x(1) = \{x \in X : u_1 \in U - D_u(x)\}.$$

If for all $i \leq p \quad u_i = \overline{u}_i$ (or $u_i = \hat{u}_i$), put $\hat{D}_x(p) = X$ (or $\overline{D}_x(p) = X$).

The successive estimation of $\overline{x}$ may be performed in a closed-loop learning system where u_i is the sequence of the decisions. The decision making algorithm is as follows:

1. Put u_n at the input of the plant and measure y_n.
2. Test the property $y_n \in D_y$, i.e. determine j_n.
3. Determine $\Delta_x(n)$ using the estimation algorithm with the knowledge validation and updating.
4. Choose randomly x_n from $\Delta_x(n)$, put x_n into $R(u, y; x)$ and determine $D_u(x)$, or put x_n directly into $D_u(x)$ if the set $D_u(x)$ may be determined from R in an analytical form.
5. Choose randomly u_{n+1} from $D_u(x_n)$.

At the beginning of the learning process u_i should be chosen randomly from U. The block scheme of the learning system is presented in Fig. 7.9 where G_1 and G_2 are the generators of random variables for the random choosing of x_n from $\Delta_x(n)$ and u_{n+1} from $D_u(x_n)$, respectively. Under some assumptions concerning the probability distributions describing the generators G_1 and G_2 it may be proved that $\Delta_x(n)$ converges to $\overline{x}$ with probability 1 [25].

The *a priori* information on the unknown parameter x in the form of the certainty distribution $h_x(x)$ given by an expert may be used in the learning system in two ways:

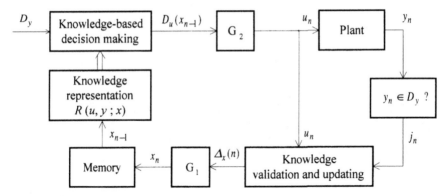

Fig. 7.9. Block scheme of learning system

1. The parameters of the distribution $h_x(x)$ may be successively adjusted in a closed-loop adaptive system.
2. The expert may change successively the form or the parameters of the distribution $h_x(x)$ using the results of current observations.

Problems concerning the application of uncertain variables in complex uncertain systems with a distributed knowledge [22] and the application of uncertain variables in learning systems with the successive knowledge updating form two main directions of further researches in the area considered in this work.

Index

References

1. Ayyub BM, Gupta MM (eds) (1994) Uncertainty Modeling and Analysis: Theory and Applications. North Holland, Amsterdam

2. Beferhat S, Dubois D, Prade H (1999) Possibilistic and standard probabilistic semantics of conditional knowledge bases. J. Logic Comput., 9:873–895

3. Bledsoe WW (1988) A survey of automated deduction. Morgan Kaufmann, San Mateo, CA

4. Bubnicki Z (1964) On the stability condition of nonlinear sampled-data systems. IEEE Trans. on AC, 9:280–281

5. Bubnicki Z (1967) On the convergence condition in discrete optimisation systems. Automat. Remote Control (Automatika i Telemekhanika), 10:115–123

6. Bubnicki Z (1968) On the linear conjecture in the deterministic and stochastic stability of discrete systems. IEEE Trans. on AC, 13:199–200

7. Bubnicki Z (1978) Two-level optimization and control of the complex of operations. In: Proceedings of VII World IFAC Congress, Vol 2. Pergamon Press, Oxford

8. Bubnicki Z (1980) Identification of Control Plants. Elsevier, Oxford – Amsterdam – New York

9. Bubnicki Z (1992) Decomposition of a system described by logical model. In: Trappl R (ed) Cybernetics and Systems Research, Vol 1. World Scientific, Singapore, pp 121–128

10. Bubnicki Z (1997) Knowledge updating in a class of knowledge-based learning control systems. Systems Science, 23:19–36

11. Bubnicki Z (1997) Logic-algebraic approach to a class of knowledge based fuzzy control systems. In: Proc. of the European Control Conference ECC 97, Vol. 1. Brussels

12. Bubnicki Z (1997) Logic-algebraic method for a class of dynamical knowledge-based systems. In: Sydow A (ed) Proc. of the 15th IMACS World Congress on Scientific Computation, Modelling and Applied Mathematics, Vol 4. Wissenschaft und Technik Verlag, Berlin, pp 101–106

13. Bubnicki Z (1997) Logic-algebraic method for a class of knowledge based systems. In: Pichler F, Moreno-Diaz R (eds) Computer Aided Systems Theory. Lecture Notes in Computer Science, Vol 1333. Springer-Verlag, Berlin, pp 420–428

14. Bubnicki Z (1998) Logic-algebraic method for knowledge-based relation systems. Systems Analysis Modelling Simulation, 33: 21–35

15. Bubnicki Z (1998) Uncertain logics, variables and systems. In: Guangquan L (ed) Proc. of the 3rd Workshop of International Institute for General Systems Studies. Tianjin People's Publishing House, Tianjin, pp 7–14

16. Bubnicki Z (1998) Uncertain variables and logic-algebraic method in knowledge based systems. In: Hamza MH (ed) Proc. of IASTED International Conference on Intelligent Systems and Control. Acta Press, Zurich, pp 135–139

17. Bubnicki Z (1998) Uncertain variables and their applications in uncertain control systems. In: Hamza MH (ed) Modelling, Identification and Control. Acta Press, Zurich, pp 305–308

18. Bubnicki Z (1999) Learning control systems with relational plants. In: Proc. of the European Control Conference ECC 99. Karlsruhe

19. Bubnicki Z (1999) Learning processes and logic-algebraic method in knowledge-based control systems. In: Tzafestas SG, Schmidt G (eds) Progress in System and Robot Analysis and Control Design. Lecture Notes in Control and Information Sciences, Vol 243. Springer Verlag, London, pp 183–194

20. Bubnicki Z (1999) Uncertain variables and learning algorithms in knowledge-based control systems. Artificial Life and Robotics, 3:155–159

21. Bubnicki Z (2000) General approach to stability and stabilization for a class of uncertain discrete non-linear systems. International Journal of Control, 73:1298-1306

22. Bubnicki Z (2000) Knowledge validation and updating in a class of uncertain distributed knowledge systems. In: Shi Z, Faltings B, Musen M (eds) Proc. of 16th IFIP World Computer Congress. Intelligent Information Processing. Publishing House of Electronics Industry, Beijing, pp 516–523

23. Bubnicki Z (2000) Learning control system for a class of production operations with parametric uncertainties. In: Groumpos PG, Tzes AP (eds) Preprints of IFAC Symposium on Manufacturing, Modeling, Management and Control. Patras, pp 228–233

24. Bubnicki Z (2000) Learning process in an expert system for job distribution in a set of parallel computers. In: Proc. of the 14th International Conference on Systems Engineering, Vol 1. Coventry, pp 78–83

25. Bubnicki Z (2000) Learning processes in a class of knowledge-based systems. Kybernetes, 29:1016–1028

26. Bubnicki Z (2000) Uncertain variables in the computer aided analysis of uncertain systems. In: Pichler F, Moreno-Diaz R, Kopacek P (eds) Computer Aided Systems Theory. Lecture Notes in Computer Science, Vol 1798. Springer-Verlag, Berlin, pp 528–542

27. Bubnicki Z (2001) A probabilistic approach to stability and stabilization of uncertain discrete systems. In: Preprints of 5th IFAC Symposium "Nonlinear Control Systems". St.-Petersburg

28. Bubnicki Z (2001) A unified approach to decision making and control in knowledge-based uncertain systems. In: Dubois Daniel M (ed) Computing Anticipatory Systems: CASYS'00 – Fourth International Conference. American Institute of Physics, Melville, N. York, pp 545–557

29. Bubnicki Z (2001) A unified approach to descriptive and prescriptive concepts in uncertain decision systems. In: Proc. of the European Control Conference ECC 01. Porto, pp 2458–2463

30. Bubnicki Z (2001) Application of uncertain variables and logics to complex intelligent systems. In: Sugisaka M, Tanaka H (eds) 2001 Proc. of the 6th International Symposium on Artificial Life and Robotics, Vol 1. Tokyo, pp 220–223

31. Bubnicki Z (2001) Application of uncertain variables to control for a class of production operations with parametric uncertainties. In: Preprints of IFAC Workshop on Manufacturing, Modelling, Management and Control. Prague, pp 29–34.

32. Bubnicki Z (2001) The application of learning algorithms and uncertain variables in the knowledge-based pattern recognition. Artificial Life and Robotics, 5: (in press)

33. Bubnicki Z (2001) Uncertain logics, variables and systems. In: Bubnicki Z, Grzech A (eds) Proc. of the 14th International Conference on Systems Science, Vol I. Wroclaw, pp 34–49.

34. Bubnicki Z (2001) Uncertain variables – a new tool for analysis and design of knowledge-based control systems. In: Hamza MH (ed) Modelling, Identification and Control, Vol II. Acta Press, Zurich, pp 928–930

35. Bubnicki Z (2001) Uncertain variables and their applications for a class of uncertain systems. International Journal of Systems Science, 32:651–659

36. Bubnicki Z (2001) Uncertain variables and their application to decision making. IEEE Trans. on SMC, Part A: Systems and Humans, 31

37. Dubois D, Prade H (1988) Possibility Theory – An Approach to the Computerized Processing of Uncertainty. Plenum Press, N. York

38. Dubois D, WellmanM P, D'Ambrosio B, Smets P (eds) (1992) Uncertainty in Artificial Intelligence. Morgan Kaufmann, San Mateo, CA

39. Kacprzyk J (1997) Multistage Fuzzy Control. Wiley, Chichester

40. Kaufmann A, Gupta MM (1985) Introduction to Fuzzy Arithmetic: Theory and Applications. Van Nostrand Reinhold, N. York

41. Klir GJ, Folger TA (1988) Fuzzy Sets, Uncertainty, and Information. Prentice-Hall, Englewood Cliffs, NJ

42. Kruse R, Schwecke E, Heinsohn J (1991) Uncertainty and Vagueness in Knowledge Based Systems: Numerical Methods. Springer-Verlag, Berlin

43. Orski D (1998) Bubnicki method for decision making in a system with hybrid knowledge representation. In: Bubnicki Z, Grzech A (eds) Proc. of the 13th International Conference on Systems Science, Vol 2. Wroclaw, pp 230–237

44. Ostrovsky GM, Volin Yu M, Senyavin MM (1997) An approach to solving the optimization problem under uncertainty. International Journal of Systems Science, 28: 379–390

45. Pozniak I (1995) Application of Bubnicki method to knowledge-based computer load sharing. In: Bubnicki Z, Grzech A (eds) Proc. of the 12th International Conference on Systems Science, Vol 3. Wroclaw, pp 290–297

46. Pozniak I (1996) Knowledge-based algorithm by using Bubnicki method to improve efficiency of parallel performing the complex computational jobs. In: Proc. of the 11th International Conference on Systems Engineering, Las Vegas, pp 817–822

47. Ranze K C, Stuckenschmidt H (1998) Modelling uncertainty in expertise. In: Cuena J (ed) Proc. of the XV. IFIP World Computer Congress, Information Technologies and Knowledge Systems. Österreichische Computer Gesellschaft, Vienna, pp 105–116

48. Rapior P (1998) The Bubnicki method in knowledge-based admission control for ATM networks. In: Bubnicki Z, Grzech A (eds) 1998 Proc. of the 13th International Conference on Systems Science, Vol 2. Wroclaw, pp 238–243

49. Szala M (2002) Two-level pattern recognition in a class of knowledge-based systems Knowledge-Based Systems, 15

50. Xu J-X, Lee TH, Jia Q-W (1997) An adaptive robust control scheme for a class of nonlinear uncertain systems. International Journal of Systems Science, 28:429–434

51. Yager RR, Kacprzyk J, Fedrizzi M (1994) Advances in the Dempster-Shafer Theory of Evidence. J. Wiley, N. York

52. Yager YY (2000) Fuzzy modeling for intelligent decision making under uncertainty. IEEE Trans. on SMC, Part B: Cybernetics, 30:60–70

53. Zadeh LA (1978) Fuzzy sets as a basis for a theory of possibility. Fuzzy Sets and Systems, 1:3–28

54. Zadeh L, Kacprzyk J (eds) (1992) Fuzzy Logic for the Management of Uncertainty. J. Wiley, N. York

55. Zimmermann HJ (1987) Fuzzy Sets, Decision Making, and Expert Systems. Kluwer, Boston

Lecture Notes in Control and Information Sciences

Edited by M. Thoma and M. Morari
1998–2002 Published Titles:

Vol. 259: Isidori, A.; Lamnabhi-Lagarrigue, F.; Respondek, W. (Eds)
Nonlinear Control in the Year 2000 Volume 2
640 pp. 2001 [1-85233-364-2]

Vol. 260: Kugi, A.
Non-linear Control Based on Physical Models
192 pp. 2001 [1-85233-329-4]

Vol. 261: Talebi, H.A.; Patel, R.V.; Khorasani, K.
Control of Flexible-link Manipulators
Using Neural Networks
168 pp. 2001 [1-85233-409-6]

Vol. 262: Dixon, W.; Dawson, D.M.; Zergeroglu, E.; Behal, A.
Nonlinear Control of Wheeled Mobile Robots
216 pp. 2001 [1-85233-414-2]

Vol. 263: Galkowski, K.
State-space Realization of Linear 2-D Systems with
Extensions to the General nD (n>2) Case
248 pp. 2001 [1-85233-410-X]

Vol. 264: Baños, A.; Lamnabhi-Lagarrigue, F.; Montoya, F.J
Advances in the Control of Nonlinear Systems
344 pp. 2001 [1-85233-378-2]

Vol. 265: Ichikawa, A.; Katayama, H.
Linear Time Varying Systems and Sampled-data Systems
376 pp. 2001 [1-85233-439-8]

Vol. 266: Stramigioli, S.
Modeling and IPC Control of Interactive Mechanical
Systems — A Coordinate-free Approach
296 pp. 2001 [1-85233-395-2]

Vol. 267: Bacciotti, A.; Rosier, L.
Liapunov Functions and Stability in Control Theory
224 pp. 2001 [1-85233-419-3]

Vol. 268: Moheimani, S.O.R. (Ed)
Perspectives in Robust Control
390 pp. 2001 [1-85233-452-5]

Vol. 269: Niculescu, S.-I.
Delay Effects on Stability
400 pp. 2001 [1-85233-291-316]

Vol. 270: Nicosia, S. et al.
RAMSETE
294 pp. 2001 [3-540-42090-8]

Vol. 271: Rus, D.; Singh, S.
Experimental Robotics VII
585 pp. 2001 [3-540-42104-1]

Vol. 272: Yang, T.
Impulsive Control Theory
363 pp. 2001 [3-540-42296-X]

Vol. 273: Colonius, F.; Grüne, L. (Eds)
Dynamics, Bifurcations, and Control
312 pp. 2002 [3-540-42560-9]

Vol. 274: Yu, X.; Xu, J.-X. (Eds)
Variable Structure Systems:
Towards the 21st Century
420 pp. 2002 [3-540-42965-4]

Vol. 275: Ishii, H.; Francis, B.A.
Limited Data Rate in Control Systems with Networks
171 pp. 2002 [3-540-43237-X]

Printing (Computer to Film): Saladruck, Berlin
Binding: Stürtz AG, Würzburg